SQL Server 2025 Unveiled

The AI-Ready Enterprise Database with Microsoft Fabric Integration

Bob Ward

Foreword by Shireesh Thota

Apress®

SQL Server 2025 Unveiled: The AI-Ready Enterprise Database with Microsoft Fabric Integration

Bob Ward
North Richland Hills, TX, USA

ISBN-13 (pbk): 979-8-8688-1846-2 ISBN-13 (electronic): 979-8-8688-1847-9
https://doi.org/10.1007/979-8-8688-1847-9

Managing Director, Apress Media LLC: Welmoed Spahr
Acquisitions Editor: Shaul Elson
Development Editor: Laura Berendson
Coordinating Editor: Gryffin Winkler

Cover image designed by fpvproductions on freepik (www.freepik.com)

Distributed to the book trade worldwide by Springer Science+Business Media New York, 1 New York Plaza, New York, NY 10004. Phone 1-800-SPRINGER, fax (201) 348-4505, e-mail orders-ny@springer-sbm.com, or visit www.springeronline.com. Apress Media, LLC is a Delaware LLC and the sole member (owner) is Springer Science + Business Media Finance Inc (SSBM Finance Inc). SSBM Finance Inc is a **Delaware** corporation.

For information on translations, please e-mail booktranslations@springernature.com; for reprint, paperback, or audio rights, please e-mail bookpermissions@springernature.com.

Apress titles may be purchased in bulk for academic, corporate, or promotional use. eBook versions and licenses are also available for most titles. For more information, reference our Print and eBook Bulk Sales web page at http://www.apress.com/bulk-sales.

Any source code or other supplementary material referenced by the author in this book is available to readers on GitHub. For more detailed information, please visit https://www.apress.com/gp/services/source-code.

If disposing of this product, please recycle the paper

This book is dedicated to all the Microsoft employees and members of the community that brought together this amazing release on such a short runway. This has been one of the best releases I've had the privilege to be a part of in my own small way.

Table of Contents

About the Author

Bob Ward is Principal Architect for the Microsoft Azure Data team, overseeing the development of Microsoft SQL from ground to cloud to fabric. With more than 32 years at Microsoft, Bob has contributed to every version of SQL Server, from OS/2 1.1 to SQL Server 2025, including Azure SQL and SQL Database in Fabric. A renowned speaker on SQL Server, Azure SQL, AI, and Microsoft Fabric, Bob frequently presents on new releases, internals, and specialized topics at events such as SQLBits, Microsoft Build, Microsoft Ignite, PASS Summit, Fabric Community Conference, and VS Live. Follow him on Twitter @bobwardms or connect on LinkedIn at linkedin.com/in/bobwardms. Bob is the author of several Apress books, including *Pro SQL Server on Linux*, *SQL Server 2019 Revealed*, *Azure SQL Revealed*, *SQL Server 2022 Revealed*, and *Azure SQL Revealed, 2nd Edition*.

About the Technical Reviewer

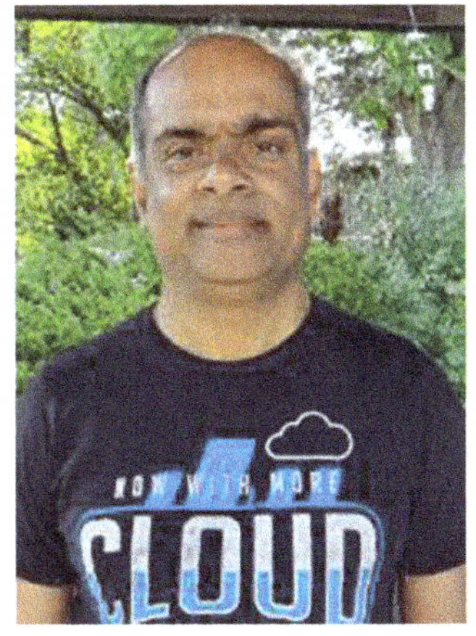

Raj Pochiraju is Principal Product Owner for SQL Server 2025 at Microsoft, bringing over 20 years of dedicated experience with the company. Throughout his career, Raj has maintained a deep focus on the SQL Server product, both on-premises and in Azure SQL cloud environments. He began his journey during the SQL Server 6.0 era, joining Microsoft as a field engineer where he specialized in troubleshooting and resolving critical issues affecting mission- and business-critical SQL Server workloads. Leveraging his extensive expertise, Raj pioneered the development of SQL Server RAP as a Service—a proactive solution designed to help customers implement best practices for SQL Server. He was also a founding member of Microsoft's flagship Database Migration Service (DMS), guiding organizations in migrating their SQL Server databases to Azure SQL. Raj has also led the development of Arc-enabled SQL Server, a technology that extends Azure services to SQL Servers running outside of Azure, empowering customers to optimize their operations and management. Raj now leads the release of SQL Server 2025, an AI-ready enterprise database.

Acknowledgments

First and always first, I want to thank God for the gift of his son, Jesus. I believe with all my heart that Jesus Christ is my savior and the redemption of my life. I'm always looking for a firm foundation, and I believe I have it in Jesus.

No one makes a bigger sacrifice when I write books than my amazing, beautiful wife, Ginger. I've been married now for over 36 years, and I am so humbled that God brought Ginger into my life. She is the best example I know for our family motto of "grace and truth." She is my partner for life, and I don't thank her enough for everything she does for me (especially those "Hey, listen what I learned today" moments).

My sons, Troy and Ryan, are the light of my life. I believe God intended for me to raise better men than myself. Each and every time I see them, I can see that I am fulfilling that promise. And to my incredible daughter-in-law Blair, who is not only an inspiration to me but also the mother of my future granddaughter Elizabeth. Elizabeth may actually be born when this book comes out, but maybe years later she will read this and understand how much she means to "Papa." I also look forward so much to having Claudia Ludden as my future daughter-in-law. She is engaged to my son Troy, and I look forward to having her as part of our family for years to come. I also want to thank my mother, Annette Gibaud, who is and will always be my "Mom" and an inspiration and example for me for my entire life.

A huge thank you to Apress. Shaul Elson has such patience! He is so professional and has balanced so well nudging me to finish this book while allowing me to ensure I create something of value and quality. I also want to thank Gryffin Winkler for all the logistics of making this book a reality. In addition, I want to personally thank Welmoed Spahr, Vice President at Springer, for all her support in publishing my books. Apress gave me a chance so many years ago, and I love that they let me write in my own conversational style.

You will see many different people from Microsoft show up in the book with quotes or mentions. There are so many people across different teams that made SQL Server 2025 possible, which has made the book possible. Here are a few though I'll call out without whom I could not have written this book.

Thanks for the leadership support I've received from Arun Ulag, Shireesh Thota (thank you, Shireesh, for writing the foreword), my current manager Priya Sathy, and the legendary Slava Oks. But I also want to call out my former managers Asad Khan and Sanjay Mishra. Without their support, we would not have launched SQL Server 2025, nor would I have authored this book. A huge special thanks to Paulette Zimmerman. She is Chief of Staff for Shireesh but also my friend and colleague at Microsoft. Paulette, thank you for everything you have done to have my back all the time.

A huge special callout to Naveen Prakash, one of the legends who runs much of our engineering team for SQL. I'm sure Naveen feels often I'm his nemesis because I push him so hard for our customers. What he doesn't know is that I have the utmost respect for his technical knowledge, his customer passion, his passion for the quality of our product, and how he takes care of his engineers. Right alongside Naveen is Alexey Eksarevskiy. I've known Alexey for many years, and he is one of the engineering leaders for whom I have such respect. Not only was he instrumental in delivering our AI features but also the overall release.

Our PM team is so talented, but I have to mention Ajay Jagannathan first. Ajay was with me way back in 2023 as we started planning this release. He is a friend and colleague that is one of the best I've known. And we were so fortunate to have Raj Pochiraju continue the torch in 2025 to make this release one of the best ever. And I want to thank Raj personally for being the technical reviewer of the book. He must have pulled some late-night hours as I cranked out chapters late in the publishing cycle. Thank you, Raj! I also want to call out Deepak Khare. Deepak is someone behind the scenes who has the thankless job of managing our release. He does this with such thoroughness and professionalism. We are lucky to have him.

So many others in our engineering teams, but I spent perhaps the most time with folks like Muazma Zahid and Davide Mauri on AI, Dimitri Furman on engine, and Nikola Zagorac on Change Event Streaming. Our marketing team as always was brilliant, but no one spent more time and energy on this release than Steven Wang. Steven showed such incredible patience with me throughout the entire release, and we would not have shipped this successfully without him. You will hear from them and others with quotes on their journey as you read the book.

Finally, all of our MVPs and community were incredibly helpful to help us test, give us feedback, and get the word out about SQL Server 2025. I've always said, without the #sqlfamily we don't have a product, and that holds true today.

Foreword

Every so often, a new release reminds us why SQL Server continues to be the backbone of data infrastructure for so many organizations. SQL Server 2025 is one of those moments. It brings together the reliability and security that enterprises expect with the flexibility and innovation that modern development demands. For developers, it opens new doors by simplifying integration with AI, cloud, and analytics platforms without introducing unnecessary complexity. For DBAs, it reinforces what has always made SQL Server dependable: strong performance, robust tooling, and a consistent management experience across environments.

This release is significant for introducing AI into our beloved relational database. It also reflects years of feedback from customers and partners. It is grounded in the belief that innovation should enhance, not disrupt, the skills and investments teams have built over time. Whether you are working on a high-throughput transactional system or building out real-time analytics, SQL Server 2025 offers capabilities that help you move forward with confidence.

Bob Ward has played a central role in this journey. I have known Bob for many years and continue to admire the combination of deep technical insight and humility he brings to every interaction. He has a rare ability to make complex database internals understandable and relevant to both developers and IT professionals. This is not just because he has worked on SQL Server for decades, but because he genuinely cares about helping others succeed with it. For SQL 2025, he has donned the role of the architect driving many of the pivotal product decisions and shaping it.

Perhaps just as important as his technical contributions is the role Bob has played in building and nurturing the SQL Server community. He has been a teacher in the truest sense of the word: sharing knowledge, guiding newcomers, and encouraging curiosity at every turn. Whether on stage at conferences, in online forums, or through blog posts or late-night troubleshooting calls, Bob has always led with empathy and clarity. His presence has helped shape not just how we understand SQL Server, but how we support each other in this shared ecosystem.

In *SQL Server 2025 Unveiled,* Bob does more than just document what is new in SQL Server 2025. He connects the technical to the practical. He gives us a roadmap for applying new capabilities to real-world problems. Most importantly, he reminds us that even as the tools evolve, the heart of relational database workloads continues to be rooted in security, resilience, performance, and developer-friendly experiences.

Shireesh Thota,
Corporate Vice President, Databases, Microsoft

Introduction

SQL Server is in my blood. I've been working now with every version of SQL Server that has shipped since SQL Server 1.0 on OS/2 with just a few diskettes. To see the product blossom to what it is today is nothing short of a dream for me. This book is a testament to that dream. To write about a product that honors the tradition of decades of work on an industry-proven database engine combined with the innovation of the latest in technology is an incredible story. A story that deserves to be told. I hope I give it justice in this book. I built this book specifically for those who already know the fundamentals of SQL Server. However, I've learned from my past books that even those who have some knowledge of SQL Server will see great value from it.

As with past books (except for *Azure SQL Revealed*, written in 2020), this book was written all over the world. Any author who does this "on the side" knows you have to find any small moment in time to author a book. So this book was written in hotel rooms, on beaches, on airplanes and in airports, in doctor office waiting rooms, on subways and trains, riding in the car while my wife drove, and any other spare moment I could squeeze in to get even the smallest of paragraph done.

The book was a journey from places like Sarasota (with my friends Tom and Janet Grubish) and Orlando, Florida; Las Vegas (multiple times); Redmond (multiple times); Seattle; Dallas; Charleston; Savannah (the home of Troy and Claudia); a cruise ship (with our close friends the Jollys); Paso Robles, California (with our friends the Peschells); San Jose, California (at the NVIDIA GTC event); Hilton Head Island; London (including trains and tubes); Abilene, Texas (including the home of Mark and Loria Beale and Brandon and Paige Beale and the famous Fairway Oaks Country Club); New York City; Genesee, Colorado (our home away from home); Vienna, Austria; and on walks around my house at Green Valley Park and the confines of my home (Ward Ranch) in North Richland Hills, Texas.

As with all of my books, the first chapter is a story of the making of this product. If that is not your thing, you can dive right into Chapter 2 to learn details of installation and upgrade. However, I have feedback that many people love to hear the "making of ..." story, so you might enjoy it. Chapter 1 also though includes an introduction into the overall release, so I feel you will want to at least browse this chapter toward the end of it.

I have examples throughout the book that you can get along with the book or at my site at **aka.ms/sql2025bookextra**, including any errata or details about SQL Server 2025 that are revealed after the writing of this book.

Chapters 3 and 4 are **all about AI**. AI is one of the biggest leaps in innovation we have made with SQL Server in years. I felt like I needed to introduce fundamental concepts of AI first in Chapter 3 and then bring in how SQL Server 2025 fits into that picture in Chapter 4. Chapter 5 is all about developers, so if you want to jump right into JSON, Regular Expressions (RegEx), T-SQL, Change Event Streaming (CES), and REST (Representational State Transfer) APIs, you will love this chapter. Then in Chapter 6, I pivot to talk about Azure Arc, especially how it lights up new security possibilities to go "passwordless" with a Managed Identity.

The "meat and potatoes" of SQL Server is the engine, and Chapter 7 will not disappoint. This is a huge chapter covering security, performance, and availability. Core engine #sqlhead lovers will love the details.

I conclude the book by talking about Fabric Mirroring for SQL Server 2025 in Chapter 8, which I think anyone should read to see what is possible with the Microsoft Fabric platform. And the final chapter is the story of Microsoft SQL, ground to cloud to fabric, so you know that SQL Server 2025 is at the heart of an amazing database platform everywhere you need it.

Authoring books is a labor of love, and I enjoyed writing this book just as much as the previous ones. Authors just don't write. We learn along the way, and I learned myself a few things I didn't know as part of this journey. When you are done with this book, my hope is you understand why we call SQL Server 2025 **the AI-ready enterprise database**. It is ready for you today!

Bob Ward, North Richland Hills, Texas
September 2025

CHAPTER 1

The AI-Ready Enterprise Database

SQL Server 2025 accelerated into one of the most amazing projects I've been a part of at Microsoft in my 32-year career. In this chapter, you will learn some fun history of the origins of SQL Server 2025 and how it came to be. You will also get an introduction into SQL Server 2025 and its major new capabilities. As part of the introduction in this chapter, I'll talk about foundations from past releases, important tools, customer stories, and a new architecture to realize with the modern innovations in SQL Server 2025. I'll close by giving you additional resources you can use on your journey to make the most of SQL Server 2025. If you want to skip the history of the making of the product, go right to the section titled "Foundations" later in this chapter.

Project Kauai Is Born

I don't get nervous many times these days working at Microsoft when presenting an idea or proposal to our leaders. But I will admit as I was waiting for my Vice President at the time, Asad Khan, to meet me in downtown Seattle during Microsoft Build in 2024, I was a bit uneasy. I was presenting at Microsoft Build that week, but instead of catching a ride to our Microsoft campus in Redmond, Washington, I had asked Asad to come to downtown Seattle to meet me, Muazma Zahid, Anna Hoffman, and my manager at the time, Sanjay Mishra.

Muazma, Ajay Jagannathan, and I had been working on a proposal for SQL Server vNext (anytime we don't have a project name, we just call the next version of SQL Server vNext). And we wanted Asad to come downtown to the Seattle Summit Convention Center for a focused meeting on such an important topic. Our meeting went better than

B. Ward, *SQL Server 2025 Unveiled*, https://doi.org/10.1007/979-8-8688-1847-9_1

I expected, but what I didn't know was that the result of this meeting would launch one of the most incredible examples of speed, focus, and teamwork I've been a part of at Microsoft in my career.

Like any major release of SQL Server, the initial planning happens almost immediately after the release of the current one. So even as early as late 2022, our team was already planning and thinking of the next version of SQL Server, right after the launch of SQL Server 2022 (and, yes, there was also a book about that release called *SQL Server 2022 Revealed*).

In these early times, there was no specific timeline set for the release, but our early thinking was probably 2026. In fact, as we moved forward during the spring and summer of 2023, some people on our team were already using the term SQL Server 2026, instead of vNext or any project name.

One of the best aspects of having Azure SQL is that it is constantly updated with features and enhancements. Therefore, planning and design are always happening for SQL Server. So even by the summer of 2023, while we were talking about what the major focus on SQL Server vNext could be, most of this conversation centered around what enhancements we were doing in Azure and how this *cloud-first* approach could be brought into the next release of SQL Server.

These initial conversations happened between me, Ajay, Naveen Prakash, Sanjay, and Asad. The consensus at this time was still likely a 2026 timeframe with a focus on our Azure SQL enhancements, ensuring we looked at top customer feedback areas, but also ensuring that we continued the success of connecting SQL Server with Azure Arc.

As we moved into the fall of 2023, I started spending more time working closely with Muazma Zahid and Davide Mauri on AI. We all knew that AI capabilities needed to be part of the Azure SQL and SQL Server story because of the generative AI explosion in the industry. The question was more what exactly we should focus on and how. It became very apparent to us, and our leaders, that two areas of investment for AI were crucial for the future of SQL–vector support and Copilot experiences.

We already started showing publicly some of these capabilities in early previews such as an early look for AI in Azure SQL. We showed an early look of Copilot at the PASS Community Summit keynote presentation in 2023 (and yet somehow I keep losing to Anna Hoffman) and vector search with Azure SQL at Microsoft Ignite 2023. Muazma and Davide specifically had been looking at how we could store embeddings from AI models in Azure using a vector data type and use T-SQL to perform similarity searches in Azure SQL Database.

Flash forward back to our meeting with Asad at Build in May 2024. Why was I a bit nervous and what was our proposal? It was to accelerate our support for AI in SQL Server and launch the next release in 2025, not 2026. The market was moving so quickly for AI, including other services at Microsoft and in the industry. By the end of our meeting, Asad agreed and went back to our engineering team to get consensus so we could formalize the plan.

This meeting would lead to the launch of our official project name, Project Kauai. Behind the scenes, the week earlier we were tossing around a project name internally. We always want a cool name. I got to pick Project Dallas for SQL Server 2022. So, to be fair, my colleague Pam Lahoud (a.k.a. @SQLGoddess) got to pick this time and had already proposed a new name–Kauai. We couldn't resist the "ai" at the end of the name, but it was more than that. Kauai is one of the most iconic places in the world (at least I think so), and we wanted a name that reflected the importance and elegance of this release. The running joke throughout the release was whether the "ship party" could be held where the project name is located.

The Move to Announce SQL Server 2025

Even though we had a project name, we decided very early to call the official release SQL Server 2025. This creates a very clear and focused mission for our team. You may have heard me say often, we will "not miss the name." This tradition carries back to SQL Server 2000 from Paul Flessner and David Campbell. At this point though we were in the summer of 2024, so if we were to call this SQL Server 2025, when would we announce it? I will admit that being one of the top leaders in our industry for SQL Server, it is tough when you are asked "When is the next version of SQL Server coming?" and you must provide an answer like "We are always looking to innovate for SQL Server but are not prepared at this time to announce anything." It was not even hard since I knew we had a plan and a mission for 2025. Luckily, Asad Khan went on record at SQLBits in March of 2024 and had said in the keynote presentation, "We will have a next version of SQL Server," which at least to our community and industry was a promise that the on-premises version of SQL Server was not going away. Around this same time, we made an organization change for the incredible Joe Sack to take over lead product management for SQL Server 2025.

We knew AI would be a major focus of new capabilities but needed to ensure we had a clear plan on what other features would make the cut for the release at General Availability (GA) (as I said earlier we are always working on new features; it was just a matter of which ones were crisp enough to make the product release). Fortunately, after so many years of cranking out SQL Server releases, we have a playbook on how to do this. In addition, we have key SQL Server veterans behind the scenes planning and executing like Deepak Khare along with Ajay to work out the release plan. We had engineering leadership from Vice President Naveen Prakash, who has a long-standing record working across SQL Server but an incredible breadth and depth of knowledge of the engine and how to deliver a release. Principal Software Engineer Alexey Eksarevskiy, with whom I've worked for years, started guiding the release from both engineering and AI perspectives.

We also spent a great deal of time in the summer of 2024 working on the AI story, our architecture, and how we would expose this to users. Security and AI model choice were #1 on our minds. Muazma, Davide Mauri, Alexey, Brian Spendolini, and I spent quite a bit of time working through this. You will see the results of this work in Chapters 3 and 4 of this book.

On the marketing side, Debbi Lyons, who has been with us for so many SQL Server releases, in addition to Azure SQL, was right with us for our marketing plans, especially on how to announce and go to market. She brought in Steven Wang to be Chief Marketing Program Manager, dedicated to SQL Server 2025. In addition, Sonya Waitman and Govanna Flores also started getting involved in our go-to-market plans. It does take a village.

As the summer moved on, our team already knew we could deliver a private preview, and we called it Community Technology Preview (CTP) 1.0 before the end of the year. This would not include all the features that would make it into the release but enough to make it compelling for customers to join our Early Adopter Program (EAP) and start kicking the tires. And we needed enough features early on that were not just about AI but other capabilities for developers and the engine. The question for us was whether we should go public and announce CTP 1.0 or wait until a public preview release. At this same time, we experienced an organization change as Sanjay Mishra left Microsoft. Sanjay was a great leader for our team and instrumental in helping us build out the plan and execute what is SQL Server 2025 today.

Maybe you have seen by now over the years that Microsoft likes to make major product announcements at big events. For SQL Server, this has generally come at four major events, Microsoft Build or SQLBits (typically in spring) or Microsoft Ignite or PASS

Data Community Summit (in the fall). We decided that if we launched a private preview of SQL Server 2025, it would be at Microsoft Ignite. After much discussion internally, we landed on the announcement of private preview for SQL Server 2025 the morning of November 19, 2024, with a blog and announcement in Scott Guthrie's keynote session at Ignite. Figure 1-1 shows the slide Scott used to make the announcement.

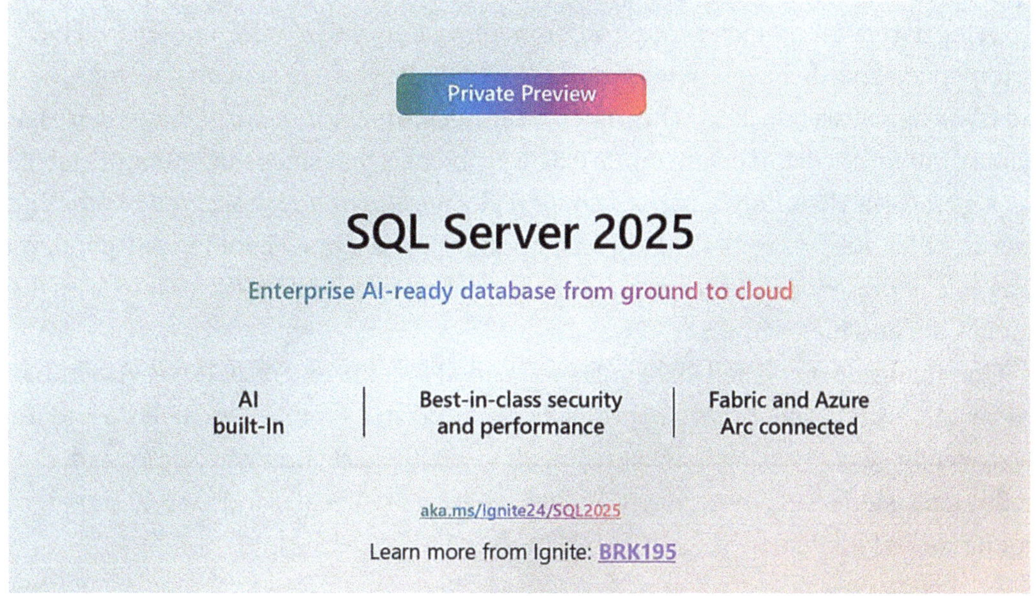

***Figure 1-1.** The announcement of the private preview for SQL Server 2025*

Asad Khan authored a blog article: "Announcing Microsoft SQL Server 2025: Enterprise AI-ready database from ground to cloud," November 19, 2024, (`https://aka.ms/ignite24/sql2025`). Figure 1-1 is actually an updated version of Scott's slide, with the BRK195 link from my Ignite breakout session added. You can still find that session at `https://youtu.be/NamOXtqDSA8`. For me personally, the morning of the launch was fun. I was actually in my hotel waiting for the launch "embargo" to be lifted and immediately posted on LinkedIn (`https://www.linkedin.com/feed/update/urn:li:activity:7264631222260105216`), X (`https://x.com/bobwardms/status/1858866303462781333`), and Reddit (`https://www.reddit.com/r/SQLServer/comments/1guxnl7/sql_server_2025_announced_at_ignite`). The Reddit post was not by me as I struggled to figure out how to post (I've got that solved now). What was incredible was the reactions: 166,000+ impressions on LinkedIn and thousands more in these other posts just after a few weeks.

As you can see from the slide, we wanted our customers to know that SQL Server 2025 would focus on AI, core engine, and cloud-connected features. What was missing from the slide was a callout for some incredible new developer features. Powered behind this was a collective set of feedback from our customers. You will soon read later in this chapter how we changed that message. It is fun to go back and see how we presented the private preview and what SQL Server 2025 has become today.

Being part of this launch at Microsoft is both exciting and a relief. I now could tell customers that we do have a new version of SQL Server, you can sign up now to try it, and it will be released in 2025. The interest was huge, and by the end of the year, we had a large number of customers in our EAP. The problem is that since our announcement was a private preview, only Microsoft could really present on the topic. And I knew there were other features we were working on for public preview that I couldn't talk about yet. I was able to present to a larger virtual user group before the end of the year to keep the momentum and message alive.

One change that affected all of us as we headed into the end of 2024 was Asad Khan leaving Microsoft. Asad was instrumental in pushing us to get this done in 2025 and was my leader for seven years. It is always difficult to get into a rhythm when your leader for this long was leaving. But I've seen a lot of change in 32 years, and the SQL Server product stops for no one.

The Acceleration to Launch Public Preview

As we started the new year of 2025, we knew we needed to

1) Keep the momentum and message going for SQL Server 2025.

2) Keep the EAP *flowing* and have customers try our private preview builds.

3) Plan for public preview.

For my part, I played two roles for the release: internally as an architect working across all the features and releases, providing guidance and advice based on my experience with the product and our customers, and externally, as I planned out presentations both virtually and at events in the spring of 2025 such as Visual Studio Live in Las Vegas, my first time at NVIDIA GTC, the MVP Summit, the Fabric Community Conference, and SQL Saturday Austin. All of these were done with our original private preview messaging. But I was able each time to add in some new *sprinkles* as we added features from January through April for CTP builds 1.1, 1.2, 1.3, 1.4, and 1.5. One of the key messages I kept landing on at these events is that AI capabilities for SQL Server 2025 work ground to cloud. The NVIDIA GTC event was especially fun as Muazma and I were able to work with NVIDIA to show our AI features using NVIDIA AI models and GPUs.

During these winter and early spring months, we made a few other changes within Microsoft. Priya Sathy stepped in as our new leader for product management for SQL, and Slava Oks became our new engineering leader. Joe Sack left Microsoft, which was one of the hardest phone calls I've had in some time. Like Asad and Sanjay, Joe was such an instrumental part of getting us to this point with SQL Server 2025. And he is also just an incredible person, but he had dreams of doing something a bit different.

Another change was Ajay moving to a new role working on Fabric Mirroring and Raj Pochiraju taking a new role for overall product manager for the release. Ajay and I had worked so long together to start Project Kauai, and with his knowledge of SQL Server and shipping a release, it was a bit difficult for me. Ajay should always know that SQL Server 2025 would not be what it is without him. But I also quickly learned how great Raj was in this new role. I've known Raj at Microsoft since before both of us joined engineering. He is the ultimate professional but also with a deep knowledge of SQL Server. His #1 job at this point? Get SQL Server 2025 shipped as a public preview. I asked Raj his perspective on the team effort it took to launch public preview. He said, "*When we kicked off SQL Server 2025, we set out with an ambitious vision—an AI-ready, enterprise-grade database delivered with urgency and precision. What made it possible was the way the team came together—across product management, engineering, marketing, and field—to move with clarity and purpose. We accelerated what would typically take years into a focused, high-impact journey and delivered a public preview that reflects both innovation and quality. It's a reflection of what's possible when a team is aligned, energized, and all-in. Along the way, we ran our largest Early Adoption Program ever, engaging hundreds of customers and MVPs across six monthly private previews. Their feedback was instrumental in shaping the product, helping us refine features and quality with every iteration.*"

Finally, Anyone Can Download It

Public preview for SQL Server is a big deal. For private preview, the world had to rely on the presentations I had been doing for public knowledge of what was in the release. EAP customers had private knowledge including documentation as we had internal PDF files for them to read and learn. When we release a public preview, anyone can download and try out SQL Server 2025 for free, and we have a complete set of public documentation. At this point all the builds were tagged CTP 1.X. Public preview for SQL Server is always CTP 2.0, and SQL Server 2025 used the same naming convention.

Besides Raj and all the incredible product managers and engineers to get the features into shape, we have other talented people to make this happen. I mentioned Deepak Khare earlier. He is the engineering release manager for SQL Server. As we moved through private preview builds and had to prepare for public preview, Deepak was in charge of our playbook having to work daily to make sure we were on track. Deepak and Raj worked together, consulting me from time to time, to get us lined up for the big announcement and public release. I asked Deepak what it is like steering such a big "ship" like SQL Server. He told me, "*Working behind the scenes on a major SQL Server release is both exhilarating and humbling. It's a world powered by meticulous planning, relentless coordination, and deep technical passion. The most inspiring part is the people. Every release is a testament to the dedication of a diverse, global team, united by a shared mission: to deliver the best possible experience for our SQL Server customers around the world. Behind every feature, every line of code, and every release note, there's a story of collaboration, innovation, and relentless pursuit of excellence.*"

Another set of incredible people work on our documentation team. Randolph West, Mike Ray, and Masha Thomas coordinated with Deepak, Raj, and our entire team to make sure the documentation was polished and ready for public consumption. They are so professional and work tirelessly to make sure the documentation is the best representation of our product.

Our marketing team with Sonya, Steven, and Govanna also kept tight track of a plan on how we would show up with go-to-market strategies and logistics. Vicki Van Damme worked with me to line up MVP and Microsoft speakers to present SQL Server 2025 at various SQL Saturday events. Our business planning team including Joe Dennis and Alicia Park were so critical in guiding our decisions across the plan for the release.

It almost feels like we are launching a mission to the moon. For the entire month of April and early May, it was like a T-minus countdown to launch every day, which we had already agreed would be May 19, 2025, at 9 AM Pacific time, as part of the Microsoft Build event. My colleague Dhananjay Mahajan (DJ) also stepped into a new role to work with our partners across SQL focusing on SQL Server 2025. In addition, our performance team led by Thierry Fevrier and Purvi Shah were working on new benchmarks with our partners with the goal to release them at General Availability. And Amit Khandelwal was working to ensure our strategy for Linux, containers, and Kubernetes was solid and lined up. All was ready to unveil SQL Server 2025 to the world.

May 2024 Events and Launch

I worked with our marketing team to secure a breakout session for Microsoft Build to give all the details of the public preview, co-speaking with Muazma. But I had a dilemma. I also needed to deliver the good news at a big event in Las Vegas, Dell Technologies World (DTW). DTW is one of the largest events in the world for on-premises customers, so naturally we needed to be there to deliver the message of SQL Server 2025. Thankfully, Aly Hirani, who leads our Microsoft partnership with Dell, was able to move my session to Monday, co-presenting with the incredible Windows expert Jeff Woolsey. In addition, our marketing team was able to move my breakout session at Microsoft Build to Tuesday (you can watch the Build session I did with Muazma at `https://build.microsoft.com/en-US/sessions/BRK207?source=/speakers/40860a91-a9fe-4ed3-b538-5888d9d6b5f3` or search BRK207 on the Build website).

But the launch was Monday morning while I was in Las Vegas. I knew I needed to repeat the same amplification of our public preview announcement like I did in November for private preview. But I also wanted to be sure to attend the Dell Technologies keynote with Michael Dell presenting. Technology to the rescue. I teed up LinkedIn, X, and Reddit posts (I just don't trust the scheduling features yet) and submitted them while watching the Dell keynote in one of the front rows (and the message from Michael Dell was "bring AI to your data," which is perfect for SQL Server 2025). The posts can be seen at my profile on these platforms:

- **LinkedIn** - `https://www.linkedin.com/posts/bobwardms_announcing-sql-server-2025-public-preview-activity-7330261199550324736-grWd`
- **X** - `https://x.com/bobwardms/status/1924495622729863225`
- **Reddit** - `https://www.reddit.com/r/SQLServer/comments/1kqfecc/announcing_the_public_preview_of_sql_server_2025`

As it was in November 2024, these posts received well over 100,000 impressions and views in only a few days. Before I flew late in the night to Seattle from Las Vegas, I had the honor of squeezing in an interview with the famous folks at theCUBE with my good friend from Dell, Rob Sonders. You can see that at `https://youtu.be/ZADA2c8-mQ8` (or just search for theCUBE and Bob Ward).

The key message for our public preview launch looked similar to private preview with a few notable changes as seen in Figure 1-2.

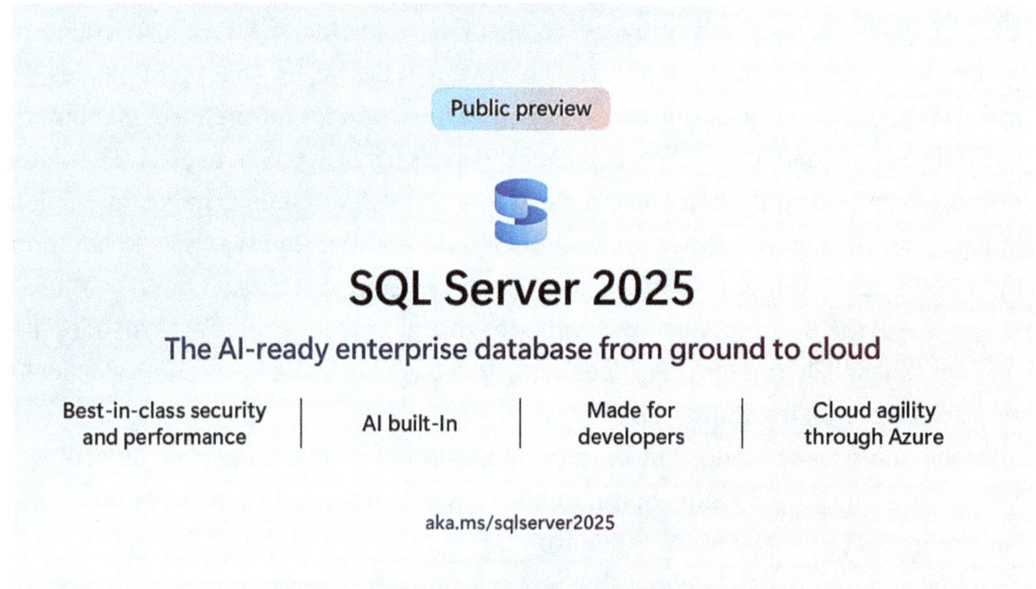

***Figure 1-2.** The launch of SQL Server 2025 Public Preview*

First, you may notice the shiny new blue icon! Yes, finally a new icon for SQL Server. It turns out Jan Nygren had already been working with us to build a possible new icon. Jan is no longer with Microsoft, but a huge callout to him for designing this sleek new look, which I believe is fitting for the innovation of SQL Server 2025. You will also notice we added a fourth "theme" to the slide called "Made for developers" paying homage to some of the most significant features for developers in the last decade.

It is worth pausing here to call out the main tag line "The AI-ready enterprise database from ground to cloud," which is also the title of this chapter. What does this really mean? First, it is calling out that SQL Server 2025 is *ready to go for AI*, built into the engine and the product. It is *enterprise* because AI can be used in a secure fashion powered by the industry-proven engine. And it is *ground to cloud*, because SQL Server 2025 can run completely on-premises, including access to AI models completely on-premises, it can be run in clouds like Azure, or it can be run on-premises and connected to Azure, with AI models like Azure AI Foundry. It can also be connected to Azure using Arc and to Fabric using Mirroring.

The link **aka.ms/sqlserver2025** on the slide points to our main blog authored by Priya Sathy talking all about what's new with the public preview.

A Rest but Momentum Still Strong

We all considered our public preview launch a success given the positive reception we received from so many. We had almost twice as many people download the public preview as compared with past releases in the timeframe right after a public preview. Music to our ears!

We didn't stop with just Dell World and Build. We launched a video from the famous Microsoft Mechanics team, which you can watch at **aka.ms/Build/sql2025mechanics**. We wrote more detailed blogs including ones from Raj and Ajay. My colleague at Microsoft Alex Powers convinced our team to host an AMA on Reddit at the SQL Server sub-reddit: `https://www.reddit.com/r/SQLServer`.

And it was not just Microsoft posting about the release. We had several prominent members of the community post content and show off features. Joey D'Antoni wrote an article in *Redmond Magazine* called "Microsoft Build: SQL Server 2025 Is (Almost) Here." Anthony Nocentino and Ben Weissman created a YouTube series called *Anthony and Ben talk about SQL Server 2025*. Daniel Hutmacher wrote a deep-dive blog article on the new JSON capabilities called "JSON indexes in SQL Server: First impressions." And post public preview, many others also started writing about the release.

As we moved toward the end of May 2025, our team was riding high knowing we had all delivered on what we intended to do over a year earlier. Now it was time to keep the momentum going over the summer and fall and push for General Availability.

Summer Momentum and General Availability

As we moved into the summer of 2025, we produced new updates and kept the word going strong at customer events and online.

CTP 2.1 and SQLBits

One of my favorite events to speak at is SQLBits. I love visiting the United Kingdom and especially London. I have been speaking at SQLBits for many years, but I knew this one would be special because the event was back in London and we had the opportunity to carry our momentum from public preview "across the pond."

The event ended up being a great success with a lot of interest in SQL Server 2025. I had the honor of presenting the SQL Server 2025 story in the keynote along with some fun interaction with the famous Patrick LeBlanc. We had just released CTP 2.1, which has some enhancements from CTP 2.0. But the big announcement Patrick made was that SQL Server Reporting Services (SSRS) was no longer going to be part of the SQL Server product moving forward to be replaced by Power BI Report Server (PBIRS). *Anyone with a paid license* of SQL Server 2025 would have license rights to install and use PBRIS.

Then I did a specific breakout session on SQL Server 2025 to a packed room. On the famous "free day," Saturday, I had the honor of co-presenting with Erin Stellato on a deep dive of the SQL Server engine. This was an important presentation because to this point I had not spent time presenting to the public more details about core engine improvements. The result was a talk called **SQL Server 2025 Engine Deep Dive**. As part of this talk, I posted these decks to a new place called **aka.ms/sqlserver2025decks**. I'm an *open-source presenter*, so I want anyone to use these for any need they have to learn or spread the goodness of SQL Server 2025. By the time you read this book, I'm sure I'll have other decks or updates on this site. When you look at these decks and see some of the slides as image in this book, you should know that I have some help on these presentations. The wonderful people at 2A Consulting often help us build these nice, polished contents with great visuals. We give them the basic slides, and they do the nice polish. I especially love working with Guy Schoonmaker on these and other Microsoft presentations. If you go to **aka.ms/sqlserver2025decks**, you will find an interesting presentation called SQL Server 2025 Community Edition. This was something fun we did at SQLBits where I invited stars from the community Deborah Melkin, Andy Yun, and Daniel Hutmacher on stage to show a five-minute demo of their favorite SQL Server 2025 feature. It is always good for someone else other than me to give the SQL Server 2025 story (and I learned some examples I had not seen before). I would also week later travel to Seattle to present SQL Server 2025 at a new event called DataCon. DJ would also that same week present our SQL Server 2025 story at HPE Discover.

I know at this point you feel that all of these presentations and events feel like the "Bob Ward show." While spreading the word about SQL Server, which is part of my job at Microsoft, there are so many behind the scenes that make all of this possible. I've pointed out some of these people in this chapter, and you will hear from others as you read other chapters in the book.

We would go on to build other updates to SQL Server in the summer, but mostly these were only for testing and feedback from our EAP customers. In the tradition of SQL Server, we then moved toward our final public preview release before GA, the Release Candidate (RC) build.

Release Candidate (RC) of SQL Server 2025

As we moved further into summer and fall, I had already secured some plans to keep presenting the message to the community, developers, and data enthusiasts. This included Visual Studio Live at the Microsoft campus, the PASS Summit Tour in New York City and the Netherlands, and the up-and-coming Fabric Community Conference in Vienna, Austria.

And in mid-August, in the tradition of past SQL Server releases, we updated SQL Server 2025 for the first Release Candidate (RC), called RC0. We also released a minor update in September called RC1. These releases marked how close we were to the launch of General Availability.

The Move to General Availability

Unfortunately, there is no way to publish a book and launch it at General Availability and write in the book the details of GA. So, as I polish the final touches of this chapter, we are still marching toward GA. The good news is there shouldn't be much difference of features or functionality at GA time that is not in this book. But there are some things I know will be announced worth noting. This includes any changes to licensing, pricing, editions, and the edition/feature matrix.

To ensure you know the latest information of anything new or different from the book, stay in touch with my updates on LinkedIn at **linkedin.com/in/bobwardms** or my posts on X with **@bobwardms** (and, yes, I'm sure I'll be somewhere at GA launch ready to hit the button to release new posts). In addition, you can also use this link for anything extra not found in the book: **aka.ms/sql2025bookextra**. At GA I'll add an addendum on this site for the book.

I hope you enjoyed taking a tour of the history and a behind-the-scenes look of how we built the product. Now let's explore some foundations and an introduction to what is in the release of SQL Server 2025.

Foundations

One time while presenting what was new for SQL Server 2019 to a customer, I talked about query store enhancements, and they said to me, "What is query store?" I paused for a second wondering how this customer who had many SQL Server deployments didn't know what query store was. I then realized it was my fault. As we talked further, turns out the latest deployment they had was SQL Server 2012. I didn't blame them for this. They were just fine with that version, but I was presenting new versions because the end of support life was close for SQL Server 2012 and they needed to upgrade. (BTW, query store first showed up in SQL Server 2016.)

From this point forward I was determined to make sure I always present the "foundations" of a new release of SQL Server, which are typically the highlights from the previous three major releases. This is because all of the releases of SQL Server are cumulative, and we may be shipping enhancements to a feature from a previous release.

Figure 1-3 represents major innovations from SQL Server 2017, 2019, and 2022.

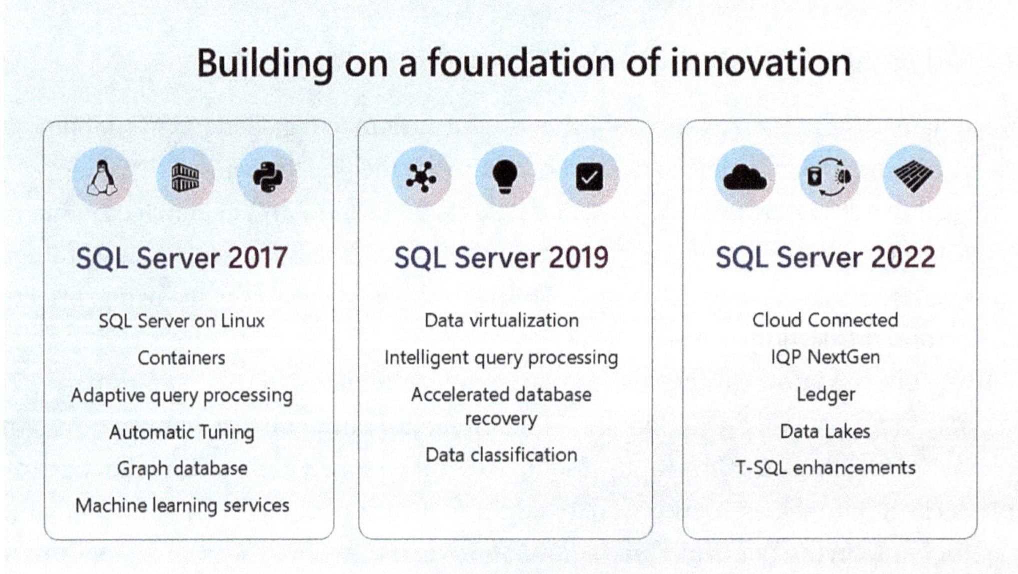

***Figure 1-3.** Innovations from previous SQL Server releases*

With that context, let's take a tour of what is new in SQL Server 2025 to set the stage for the rest of the book.

Introducing SQL Server 2025

In the past few releases of SQL Server, I tried to build a single slide that visually represented what was new about the release. I call this my "camera slide," as when I usually display it, audiences want to take a photo of it. SQL Server 2025 is no different, and this is a good way to introduce you to the new capabilities of SQL Server. It also sets the stage to introduce the main topics of the rest of the chapters of this book. Figure 1-4 is the SQL Server 2025 "camera slide."

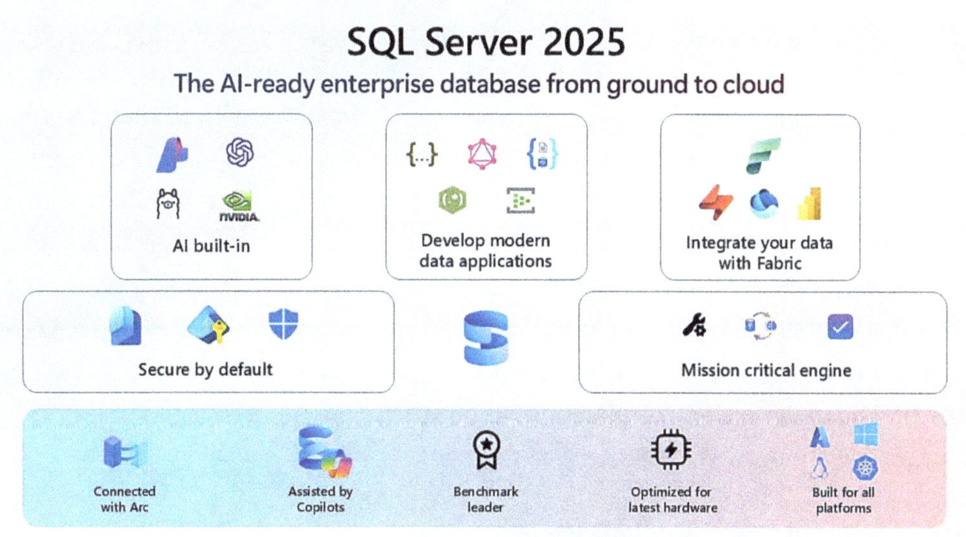

Figure 1-4. *A visual of SQL Server 2025*

Let's take a look at what is on this image to get a better feel for what is new in SQL Server 2025.

Areas of Innovation

We have introduced features in three new areas I call the *areas of innovation*. This includes features for the following.

AI Built-In

This is the *flagship* innovation for SQL Server 2025. This includes built-in support for smarter searching with vectors secured by SQL, using the familiar T-SQL language, and allows you to use the AI models of your choice ground or cloud. You will learn more about AI capabilities for SQL Server 2025 in Chapters 3 and 4 of the book.

Develop Modern Data Applications

In my opinion, this is the most significant release for developers since SQL Server 2016. Developers can use Data API Builder (DAB) to access the database through GraphQL or REST. SQL Server now has a native JSON data type, which includes an index and new T-SQL JSON functions. The T-SQL language gets some love with new and updated functions including Regular Expression (RegEx) support. In addition, we now support complete streaming of changes from the transaction log with a feature called Change Event Streaming (CES), offering a new and exciting alternative to Change Data Capture (CDC). Finally, consider a new option to call REST API endpoints running ground or cloud using a system stored procedure. This opens up new opportunities to use your data and bring in data into SQL Server from software services all through the engine without having to use a client application. I know it is a lot, but it's all goodness. And you will learn more about this in Chapter 5 of the book.

Integrate Your Data with Fabric

Microsoft Fabric (a.k.a. Fabric) has caught the industry by storm providing a new, unique unified data platform. We want every SQL user to be able to integrate their data with Fabric. One method to do this is with Mirroring SQL Server with Fabric. You will learn in Chapter 8 of the book more about Mirroring.

Meat and Potatoes

Conor Cunningham has always called the database engine the *meat and potatoes* of SQL Server. And in SQL Server 2025, we have well over *40* new features across security, performance, and availability. All of this provides a secure, mission-critical engine. You will learn about some security features in Chapter 6 of the book related to Azure Arc, but also more security features built into the engine in Chapter 7 of the book. You will also learn more about performance and availability features in Chapter 7 of the book, many based on popular customer requests.

Powering SQL Server

At the bottom of Figure 1-4, you can see a list of icons representing core foundational capabilities and concepts that power SQL Server to do more. This includes the following.

Connected with Arc

We have supported hybrid scenarios for SQL Server using Azure Arc for several releases. The value of using Arc is as it was from previous releases with one significant enhancement. That is a broader support for Microsoft Entra managed identities for both inbound (authentication) and outbound (working with an Azure resource from within the engine). I've still seen customers not understand how Azure Arc works or its value. Not your fault, but I'll fix that. Therefore, I've included coverage of this topic in its own chapter in Chapter 6 of the book.

Assisted by Copilots

Copilot experiences are providing AI assistance almost everywhere a Microsoft product or service exists. Our new SQL Server Management Studio (SSMS) has now a Copilot-assisted capability. You will learn more about this in the next section of this chapter, and you will also see it used in other chapters in this book.

Benchmark Leader

SQL Server has been the leader in TPC-E and TPC-H benchmarks working with our OEM partners for some time. Since the beginning of SQL Server 2025, we have been working on releasing new benchmark results that continue our leadership in both performance and price/performance. We typically work with our partners to publish these benchmarks at General Availability. You can always follow the latest TPC published results at `https://tpc.org`.

Optimized for Latest Hardware

We are always looking to ensure our engine is optimized for the latest hardware innovations. We typically prove this with our benchmarks, but there are also other plans you should see by the time our product is generally available.

Built for All Platforms

Don't forget that SQL Server runs on any platform you need including Windows, Linux, containers, Kubernetes, and public clouds like Azure Virtual Machines. The examples in this book will all use Windows, but I'll talk in the book about some compelling scenarios to use containers, Kubernetes, and a few improvements specific to Linux.

How important is SQL Server 2025 to the industry, community, and Microsoft? I asked Priya Sathy, Partner Director for SQL at Microsoft, for her perspective: "*SQL Server 2025 represents a major leap forward for the data platform. What excites me most is how we've embedded AI and developer capabilities directly into the SQL engine itself. This unlocks new possibilities for customers to build intelligent, real-time applications without leaving the database—whether on-premises, in Azure, or across clouds. For the industry, it's a signal that the future of data is not just about storage and performance—it's about insight and innovation, with the security guarantees SQL is known for. For the community, it's our continued commitment to SQL. And for Microsoft, it's another proud step forward in making AI truly accessible to every developer.*"

A New Wave of Tools

With the timing of the planning and execution of SQL Server 2025 came a new set of tools updated for the modern data professional. I'll call them out here briefly since I'll be using these tools, especially the new SQL Server Management Studio (SSMS), in many of the example scripts for this book.

The New SSMS

Over the last year we have shown the world that not only is SQL Server Management Studio (SSMS) still alive, but it has also almost had a complete makeover. That starts with the new version SSMS 21.

Figure 1-5 shows what is new for SSMS 21 (with its own new icon!).

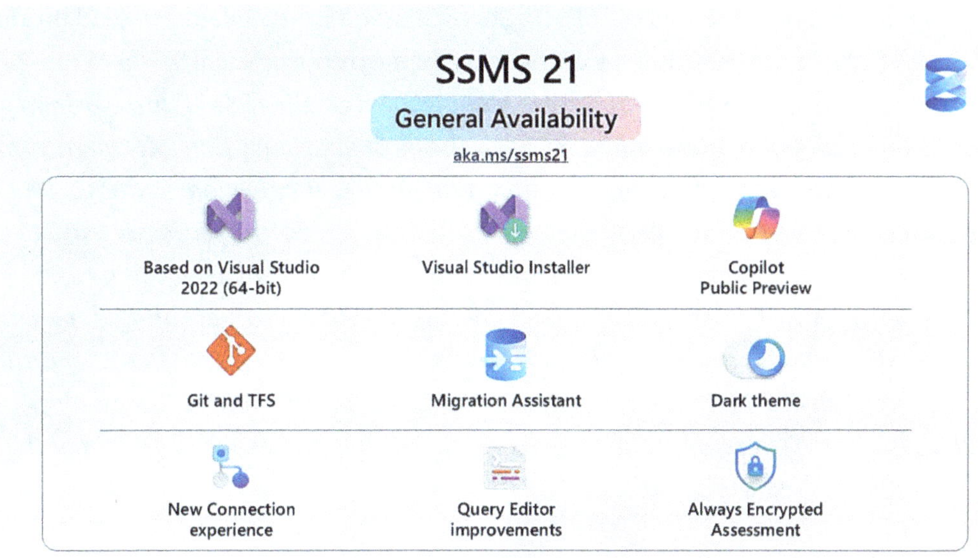

Figure 1-5. *The new SSMS 21*

If you look at this slide, you will see some incredible innovations including a completely new shell and installer experience based on Visual Studio 2022. This now enables a 64-bit version of the tool. Looking at the other items in this list, ones that stand out for me are a Copilot, Git support, query editor improvements, and (drum roll please) **dark theme**. There are many other great enhancements, which you can read more about at `https://aka.ms/ssms`.

Copilot is an interesting option for SSMS. The intention of our team is to provide AI assistance in the context of your connection to SQL Server (or Azure SQL). This could be helping you build T-SQL queries or assisting with the management of your SQL Server and databases. At the time of public preview, Copilot did require a connection to Azure and a deployment of an Azure AI Foundry model. Stay in touch with **aka.ms/ssmscopilot** for example prompts and best practices for how to use it.

Note Talk about innovation. During the end of the writing of this book, we released a new version of SSMS, **SSMS 22**, which now includes an integration with GitHub Copilot. Keep up with the latest on SSMS at `https://learn.microsoft.com/ssms/roadmap`.

SSMS is a key tool to the success of SQL Server. I asked Erin Stellato, the lead product manager for SSMS and SQL Copilot experiences, on the importance of SSMS to the future of SQL Server. Erin said, *"SQL Server Management Studio (SSMS) is the primary application for millions of users, and it provides the most direct and tangible interaction users have with SQL Server, creating the foundation for how they explore, manage, and understand the product. SSMS has grown alongside SQL Server, and the latest release reflects a renewed investment in SSMS and the data professionals who rely on it every day. With support for SQL Server 2025, users are empowered to evaluate and adopt new features; SSMS continues to serve as the bridge between the evolving data landscape and the proven strength of the SQL Server platform. Longevity isn't just about legacy—it's about relevance. As SQL Server continues to evolve, SSMS is positioned to remain at the heart of its story for years to come."*

Visual Studio Code and the mssql Extension

One of the tools we have increased our investments is for developers of SQL Server with Visual Studio Code (VSCode) and the **mssql** extension.

Here is a quick look at VSCode using the mssql extension connected to a SQL Server 2025 instance with the new GitHub Copilot experience in Figure 1-6.

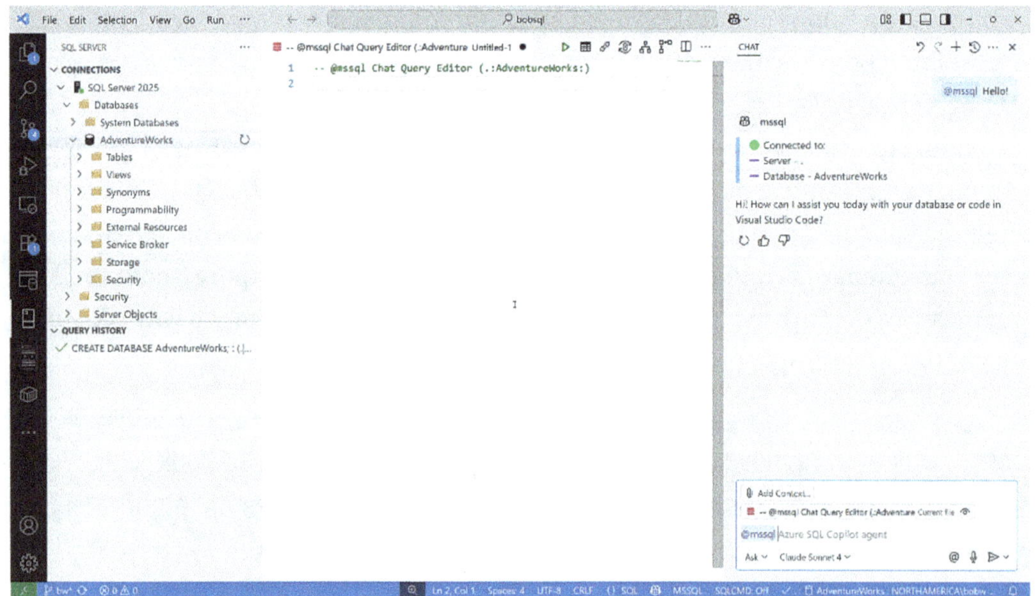

Figure 1-6. VSCode and the mssql extension with GitHub Copilot

We announced the retirement of Azure Data Studio (ADS) in February of 2025. It is our intention to replace features from that tool either in the mssql extension or in SSMS. I'll use the mssql extension in a few examples in this book. If you want to dive in now, visit **aka.ms/vscode-mssql** where you can learn about the extension and look at the GitHub repository, managed by my colleague Carlos Robles.

The SQL Server 2025 Platform Architecture

I often take walks at a park near my home called Green Valley Park, sometimes in the evenings, mornings, or even breaks in the afternoon when I work from home. I usually put on my headphones with my favorite music playlists and walk sometimes a few miles. This is a time to decompress but in many cases a time for me to innovate and think about new ideas. One day early in 2025, I thought to myself on one of these walks, *Why not build out a new architecture based on all the innovations landing in SQL Server 2025?* The idea would be to show visually to our customers what is possible now with SQL Server 2025. The result you see in Figure 1-7.

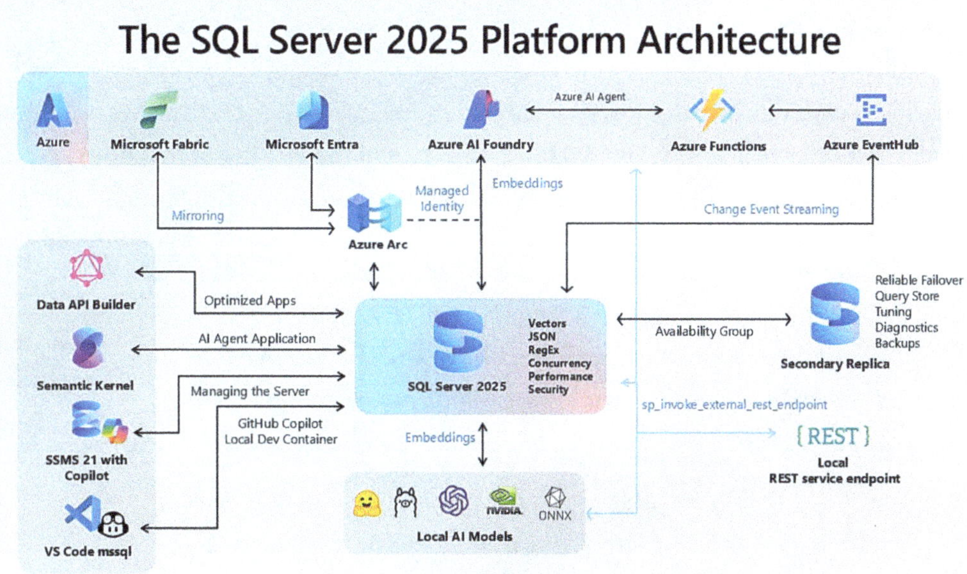

***Figure 1-7.** The SQL Server 2025 Platform Architecture*

Consider what is now possible with SQL Server 2025. You have core engine features and enhancements including specific features designed for Always On Availability Groups (AGs). You can now connect with AI models either on-premises or in clouds like Azure. You can build new data-driven applications inside the engine using REST API support. You can enhance your security with Microsoft Entra and managed identities powered by the hybrid management of Azure Arc.

You can mirror your data into the new unified data platform, Microsoft Fabric. You can build event-driven microservices including AI Agents using Change Event Streaming (CES). And you have access to build new modern AI applications with Data API Builder, Semantic Kernel, VSCode, and GitHub Copilot. And you have new management experiences assisted by Copilots with SSMS. This is both the mission-critical SQL Server you have come to love and can trust with new and modern, innovative features and integrations to allow you to go further than ever before.

This architecture is important to consider as you read other chapters in the book. Keep this picture in mind or reference back to it to see what is possible by putting many of these new capabilities together.

Customer Stories

Very early on with our private preview, we had many customers test and start seeing what was possible with SQL Server 2025. At the time of the writing of this book, I had already seen some incredible examples of customer quotes including the following:

> **Kramer and Crew** (`https://www.kramerundcrew.de/de`) told us, "*With the new semantic search and RAG capabilities in SQL Server 2025, we can empower existing GenAI solutions with data embeddings to create next-generation, more intelligent AI applications.*"
>
> **MSC** (`https://www.msc.com`), a long-time user of SQL Server, said, "*Change Event Streaming and Fabric Mirroring for SQL Server 2025 help MSC to build the bridge to bring our operational data into Microsoft Fabric.*"

> **Saxo** (`https://www.home.saxo`), another company with SQL veterans, said, "*We are especially looking forward to the performance benefits of optimized locking, new ZSTD backup compression algorithm; the enhanced reliability of availability groups; and the continued investments in security.*"

Some of these features these great companies had already looked at you have not heard about yet. But you will in subsequent chapters in this book. These customer quotes were just the start. By GA, I expect many more.

The interest in SQL Server 2025 was fulfilling and exciting for me. After our public preview launch, we started seeing the community and experts post their own experiences ranging from AI to the core engine. Just use your favorite AI tool or search engine for *SQL Server 2025*, and you will see it was not just Microsoft touting this new and exciting release.

Resources for You

As you read this book, you will see me point to various specific references for you to learn further, but I wanted to show you overall resources that can help you on your SQL Server 2025 journey:

> **aka.ms/sql2025bookextra** – I mentioned this earlier in the chapter. This is a way to stay up with any news about SQL Server 2025 that didn't make the book or errata discovered after publishing. I'll also publish all the demo scripts here, but they are also available from Apress.
>
> **aka.ms/getsqlserver2025** – At the time of the writing of this book, this points to the download for evaluation of public preview. At GA we will redirect this to the download center for Developer, Evaluation, and SQL Server Express editions.
>
> **aka.ms/sqlserver2025docs** – This points to the latest "What's new in SQL Server 2025 docs." I frequently use this link to double-check references.

aka.ms/sqlserver2025decks - This is where I put all my public presentations (demos included) for SQL Server 2025 for anyone to use. What is nice about these decks is you can see a video of some of the examples you will find in the book.

aka.ms/sqlserver2025demos - This is my GitHub repo where I put demo scripts. You will also see many of these in this book, but I may have some here not in the book, or I may provide more details in the book.

aka.ms/Build/sql2025mechanics - A video from Microsoft Mechanics showing SQL Server 2025. Always a professional and polished video with that team.

aka.ms/SQL2025videos - This is a series of videos for the Data Exposed show for SQL Server 2025. Watch the famous Anna Hoffman interview many of us on the team.

aka.ms/sqlchannel - This is the official YouTube channel for SQL Server. We have not kept this up to date, but it is our intention to start doing more here including SQL Server 2025 and SSMS.

aka.ms/sqlserverblog - This a series of technical blogs by our team including ones on SQL Server 2025.

aka.ms/sql2025blogs - This is a blogging series specifically targeting SQL Server 2025.

Tradition Meets Innovation

If you are an expert SQL Server user, you may want to just download SQL Server, install it, and dive right in. The next chapter shows you the basics to get going fast but also more information about setup, upgrade, and configuration.

Then the choice is yours to learn all about AI in Chapters 3 and 4 or find the chapter of your liking to focus on developer features, engine, Arc, or Microsoft Fabric integration. However you choose to learn moving forward, I think you will see how SQL Server 2025 embraces the tradition of an industry-proven product with the innovation you need to keep up with the modern demands of data, AI, and applications.

CHAPTER 2

Ready, Set, Go

Cameron Battagler, Principal Data and AI Technical Specialist at Microsoft, posted this comment on LinkedIn (see his post at `https://www.linkedin.com/posts/cameronbattagler_sqlserver-sql2008-sql2025-activity-7331431067788226560-pF0d`) shortly after we announced public preview: *"I know this may not bring comfort to anyone else. But the fact that this screen still looks almost exactly the same as it did with SQL 2008 nearly 20 years later really helps me deal with the constant onslaught of change."* This quote sums up why this chapter is one of the shortest in the book.

Who Should Read This Chapter?

I will assume as an author that most readers are familiar with the basics of installing SQL Server. Having said that, I cover in this chapter topics like installation fundamentals, feature differences, discontinued features, what is different in SQL Server 2025 from SQL Server 2022, deploying on other platforms, side-by-side and multi-instance deployments, upgrades, dbcompat, and configuration. If you read Chapter 2 of *SQL Server 2022 Revealed,* this chapter will look very familiar. Let me call out a few concepts I recommend you look at:

1. I moved the topic of setup and configuration of the Azure Arc extension of SQL Server to its own chapter, which is Chapter 6 of the book.
2. There are feature differences including discontinued or deprecated features from SQL Server 2022 you should review. I've covered these in the next section of the chapter. SQL Server 2022 itself had differences from previous releases, which you can read at `https://aka.ms/sqlserver2022editions`.

B. Ward, *SQL Server 2025 Unveiled,* https://doi.org/10.1007/979-8-8688-1847-9_2

3. Any details on SQL Server 2025 editions including features and limits are not announced until GA so they are not included in this book. Stay in touch with `https://aka.ms/sqlserver2025editions` for the details. I'll also post more about this also at **aka.ms/sql2025bookextra**.

4. There are a few new options for server and database configuration, which I've also covered in this chapter.

5. Always check our release notes as we can in an agile fashion add things here that are not in the standard documentation: `https://learn.microsoft.com/sql/sql-server/sql-server-2025-release-notes`.

How to Install SQL Server 2025

You may be used to getting started with SQL Server by installing the Developer or Evaluation free editions of SQL Server and testing the product. SQL Server 2025 supports both of these editions, so if you just want to jump in, go to `https://aka.ms/getsqlserver2025` and start testing! **SQL Server 2025 will be known as a new major version number of 17.X** based on the system variable, @@VERSION, and other places in the product such as the ERRORLOG.

Note SQL Server 2025 has a new option for Developer Edition to help developers choose their production target deployment, whether it be for Enterprise or Standard Edition. You can learn more about this new option in Chapter 5 of the book.

I know that some people want to know more about the installation of SQL Server before they go "straight for the bits." In this section of the chapter, I'll review the prerequisites to install SQL Server and discuss the differences from previous versions, discontinued features, and deployment options.

This section of the chapter focuses on the installation of SQL Server on Windows. The section titled "Deploying on Other Platforms" later in this chapter discusses installation on Linux, containers, Kubernetes, and Azure.

Prerequisites

The prerequisites, including resources required, to install SQL Server 2025 on Windows have not changed from previous major versions of SQL Server, except what specific Windows operating system versions or Linux distributions we support. As in the past, we will support versions of Windows Server and Windows (client) that are considered "officially" supported including any specific updates of Windows required to be supported. Consult the SQL Server 2025 release notes for any possible changes to prerequisites for SQL Server 2025 at `https://learn.microsoft.com/sql/sql-server/sql-server-2025-release-notes`.

One difference for prerequisites from previous versions of SQL Server, but is the same as SQL Server 2022, is the requirement to be connected to Azure to support some of the new *cloud-connected* scenarios. The most important requirement is that if you choose to set up Azure-connected scenarios during installation, you will need an Azure subscription and an Azure login or service principal. You do not have to set up Azure Arc during setup, but the default its "on," so you will need to "opt out" by selecting Skip on that screen. You can always go back and set it up later.

To find all the exact resource requirements you need to install SQL Server, check out documentation at `https://aka.ms/deploysqlserver2025`.

Note We have received a tremendous amount of interest for SQL Server 2025 to support ARM64 platforms. Our team has had many conversations to see how to make this happen. However, at the time of the writing of this book, we had not committed to officially supporting ARM64 for SQL Server. We may add a warning to SQL Server 2025 setup to caution you if we detect you are using a machine with an ARM64 processor.

What Is Different for SQL Server 2025?

Installing SQL Server 2025 on Windows using the "setup wizard" is remarkably similar to installing SQL Server from previous releases. Therefore, I won't show you a "screen-by-screen" experience. See the complete set of steps at `https://aka.ms/deploysqlserver2025`.

Rather, I'll discuss differences from SQL Server 2022. Let's look at each of these differences from the perspective of running SQL Server setup.

Feature Differences and Discontinued/Deprecated Features

SQL Server 2022 had several differences from previous versions on the setup choices for *features*. This included removal of features like open source packages such as R, Python, and Java (but you can install them yourself). It also included the removal of PolyBase Hadoop connectivity with Java (but PolyBase is still supported in other ways), Machine Learning Server, and Distributed Replay.

Note One nice change in SQL Server 2025 is that you don't have to install PolyBase Services as a feature to use external data sources for PARQUET or DELTA.

We have removed or discontinued the following features for SQL Server 2025:

- **Data Quality Services** (DQS) and **Master Data Services** (MDS) are discontinued and not part of SQL Server 2025. They are still available in previous versions. Therefore, you will not see these features available when you install SQL Server 2025.
- The **Hot Add CPU** feature is deprecated and may be removed in a future version.
- **Synapse Link (aka.ms/sqlsynapselink)**, which was introduced in SQL Server 2022, is discontinued and not available for SQL Server 2025. Fabric Mirroring, which is discussed in Chapter 9 of the book, is the best replacement for Synapse Link.
- **Purview access policies**, which were introduced in SQL Server 2022, are discontinued in SQL Server 2025. The recommendation is to now use *Fixed Server Roles*, which were also introduced in SQL Server 2022.

The most significant change for SQL Server 2025 is the replacement of SQL Server Reporting Services (SSRS) with Power BI Report Server (PBRIS). PBRIS is an on-premises reporting platform that offers more features than SSRS and is more compatible with Power BI. Like SSRS, PBRIS is a different setup experience than SQL Server. PBRIS was previously available to only certain licenses for SQL Server. In SQL Server 2025, *any paid edition* of SQL Server 2025 has the right to deploy PBRIS. SQL Server setup will now redirect users to the PBRIS documentation page to install PBRIS. You can read the Frequently Asked Questions (FAQ) about this announcement at `https://learn.microsoft.com/sql/reporting-services/reporting-services-consolidation-faq`. Figure 2-1 shows the new link to install PBRIS.

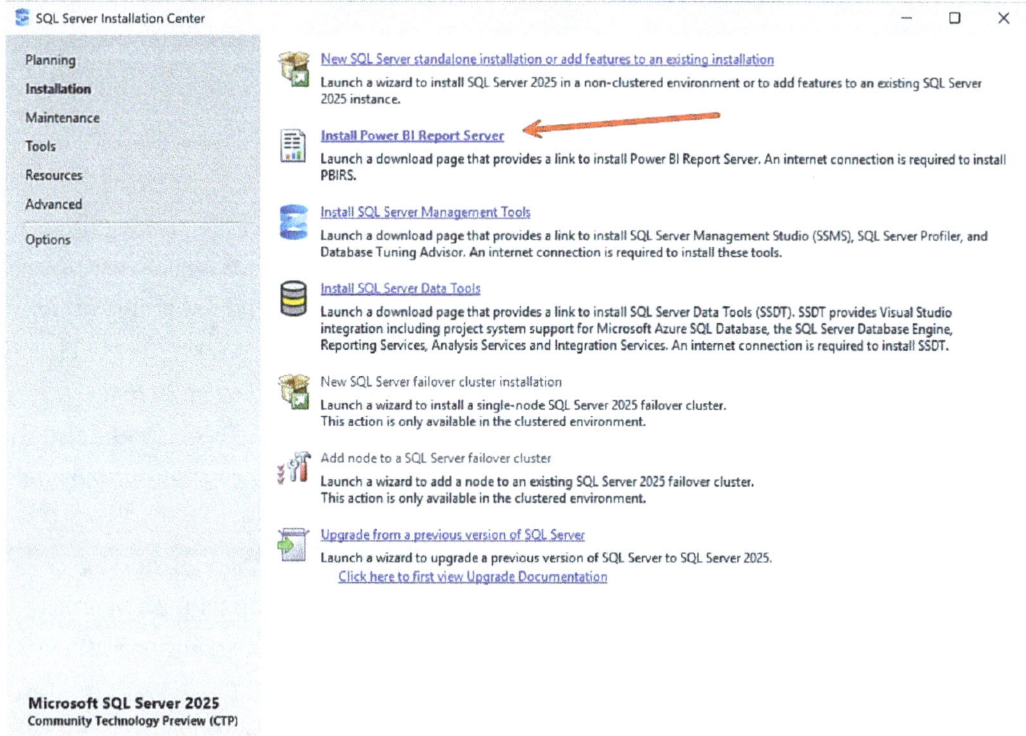

***Figure 2-1.** Link to install Power BI Report Server*

One last comment about other services:

- SQL Server Analysis Services (SSAS) has enhancements for SQL Server 2025 that you can read about at `https://learn.microsoft.com/analysis-services/what-s-new-in-sql-server-analysis-services`.
- SQL Server Integration Services (SSIS) has some features that are being marked deprecated, but the service is still supported in SQL Server 2025. You can read more at `https://learn.microsoft.com/sql/integration-services/what-s-new-in-integration-services-in-sql-server-2025`.

- One new feature name change. Machine Learning Services is now called **AI Services and extensions feature.** This feature can still be used for all the capabilities you have had with R, Python, and Java. But in addition, you will now be able to host AI models using the Open Neural Network Exchange (ONNX) runtime. You will learn about this in Chapter 4 of the book.

Other Installation Methods

SQL Server setup for Windows still supports options without a user interface through the command line. You can find examples of how to do this and see all the options at `https://docs.microsoft.com/sql/database-engine/install-windows/install-sql-server-from-the-command-prompt`. The only differences for SQL Server 2025 are the removal of options for discontinued features as described earlier in this chapter and the addition of new parameters to support the Azure extension for SQL Server (which was also there in SQL Server 2022).

If you are new to setting up SQL Server, you should know that the SQL Server Evaluation and Developer editions come with an "easy setup" mode with an installer that installs just the defaults without going through any screens. If you choose this method, the Azure extension for SQL Server feature is not selected. If you want the full setup experience with the Evaluation or Developer editions, choose the *Custom* option from the initial setup screen.

Deploying on Other Platforms

This chapter has focused more on the experience of installing SQL Server on Windows. SQL Server is supported to install on other platforms including Linux, containers, and Kubernetes. Here are resources to get you started to deploy using these other options:

SQL Server on Linux - `https://aka.ms/sqllinux`
SQL Server containers - `https://aka.ms/sqlcontainers`
SQL Server on Kubernetes - `https://aka.ms/sqlk8s`

Note This page contains instructions on deploying a SQL container on Azure Kubernetes Service (AKS) or Red Hat OpenShift. Consult your Kubernetes platform documentation on the proper method to deploy pods and containers.

Be sure to consult the latest release notes for SQL Server 2025 on Linux and containers at `https://learn.microsoft.com/sql/linux/sql-server-linux-release-notes-2025`. This includes a list of official supported Linux distributions and versions supported.

One new capability that is not specifically tied to SQL Server 2025, but the tool enhancement has landed as I'm writing this book, is the ability to deploy and configure a SQL Server container with the **mssql extension in Visual Studio Code**. Try it yourself at `https://learn.microsoft.com/sql/tools/visual-studio-code-extensions/mssql/mssql-local-container`.

In addition, SQL Server 2025 can be deployed on Azure Virtual Machines. One of the nice benefits of using Azure Virtual Machines is a pre-deployed setup of SQL Server and assisted experiences to optimize things like storage in Azure. To get more information on SQL Server on Azure Virtual Machines, consult our documentation at `https://aka.ms/sqlazurevm`.

Note SQL Server is also supported in other clouds. I would like to see you consider Azure if you want to run SQL Server in a cloud platform. But if you choose to run in another cloud, I would love you to choose SQL. Read more at `https://aws.amazon.com/microsoft/sql-server/` or `https://cloud.google.com/sql/docs/sqlserver`.

Side-by-Side and Multi-instance Installations

Like previous releases of SQL Server, SQL Server 2025 supports both side-by-side installations on the same computer with different versions (supported versions) and multiple instances (a.k.a. named instances). There are no differences in support for both side-by-side and multi-instance from previous SQL Server releases. For more information consult our documentation at `https://docs.microsoft.com/sql/sql-server/install/work-with-multiple-versions-and-instances-of-sql-server`. SQL Server on Linux supports multiple instances by using containers.

SQL Server also continues to support installations for Always On Failover Cluster Instances (FCIs) and Availability Groups.

How to Upgrade to SQL Server 2025

SQL Server 2025 supports upgrades from previous releases of SQL Server with the same methods as in previous versions, which include in-place upgrades and database restores. You can read more at `https://learn.microsoft.com/sql/database-engine/install-windows/upgrade-sql-server` for all the details.

Note At the time of the writing of this book, we had just launched a migration assistant inside SSMS (see the announcement at `https://techcommunity.microsoft.com/blog/microsoftdatamigration/general-availability-of-sql-server-migration-component-in-ssms-21/4415574`). It is our intention to include in this feature the ability to assess an upgrade to SQL Server 2025.

The Importance of dbcompat

Over my career at Microsoft, one of the reasons customers do not upgrade to a new major version of SQL Server is that the latest version is not compatible with their application.

For the past several SQL Server versions, we have been trying to convince customers and developers to test for compatibility with the database compatibility level (dbcompat) of user databases. If you upgrade the SQL Server database to a new version, we will maintain the previous dbcompat. You can read more on why this could be a new strategy for compatibility for you at `https://aka.ms/dbcompat`.

SQL Server 2025 introduces a new dbcompat level of 170 (the version of SQL Server is 17.X, hence the 17). Some features may only be available if this dbcompat is configured for your database either to preserve backward compatibility (new keyword) or to enable a new feature. I'll call these out when necessary in the book. It is our plan to support dbcompat levels as far back as 100 (SQL Server 2008) in SQL Server 2025.

Configuration

Even though this chapter is specifically designed to give you guidance on installation and upgrades, the configuration of SQL Server post installation is an important topic. But there are very few changes to SQL Server instance configuration in SQL Server 2025. There are a few new server-level **sp_configure** options. They will be discussed in this book as they relate to new features for SQL Server 2025.

In addition, there are a few new ALTER DATABASE SCOPED CONFIGURATION options, which you will read about in the book as any feature that requires a new configuration option is discussed.

There is one significant ALTER DATABASE SCOPED CONFIGURATION option that will enable some features but is general in nature and new for SQL Server 2025. This option, called **PREVIEW_FEATURES,** will be used to enable specific features at General Availability in a "preview mode." The concept is to allow developers to use a feature in "preview" even when the product is at General Availability. Feedback will allow our team to make fixes and provide enhancements. Then in a subsequent Cumulative update (CU) for SQL Server 2025, the feature will no longer need this option and become "Generally Available." You will learn more about this option as it pertains to specific features that require it later in the book. But for now you can read more about the preview feature concept at `https://learn.microsoft.com/sql/sql-server/sql-server-2025-release-notes?view=sql-server-ver17#preview-features`.

Easy to Install and Upgrade

SQL Server has been known for some time as a simple install and upgrade experience, and none of that changes in SQL Server 2025. As part of that experience, we have provided an integrated setup to get connected to Azure if that is part of your hybrid strategy.

If you read over all the details of this chapter, you are ready to dive in and learn about features. So let's roll! The next chapter is your path to get started to learn the fundamentals of AI to help go forward in Chapter 4 to build AI applications with SQL Server.

CHAPTER 3

AI Fundamentals

You might remember in Chapter 1 I described my beginnings to work on SQL and AI started in 2023. There is a bit more of the backstory for this. I remember talking to Asad Khan, our VP of SQL at the time, in the summer of 2023. He said, "Bob, please get engaged with AI." I didn't ask why because we all knew AI was coming in fast and hot to the industry. What I didn't know is exactly *what* I should do. As we approached Microsoft Ignite (and the PASS Summit) in November of 2023, it became very clear to me that we were not prepared to showcase generative AI technology with SQL. At the same time, my colleagues Muazma Zahid and Davide Mauri were already looking at vectors and embeddings with Azure SQL.

Therefore, we did two things: (1) We assembled a prototype of a vector search using SQL as a target source and showed this at Microsoft Ignite. (2) We showed a prototype of a Copilot for Azure SQL in the PASS keynote (yes, and once again I lost to Anna Hoffman in a competition—just couldn't beat that Copilot). And with any new technology, I needed to get involved, and I poured myself into the fundamentals—in this case the fundamentals of AI. That is what this chapter is all about. I will spend time in this chapter transferring to you my knowledge about AI fundamentals as a foundation to learn more about who, how, and why so you can dive into further why SQL Server 2025 has "AI built-in."

It could be you know a lot about AI fundamentals including types of AI models, vectors, embeddings, vector search, Retrieval Augmented Generation (RAG), AI tools, and Model Context Protocol (MCP). Yes, I know, it's a lot. I believe even if you have a firm grasp of today's AI technologies, you will benefit from this chapter before moving to Chapter 4.

There are no scripts in this chapter, but I have some examples I'll show you, which include the use of AI models hosted by Ollama, which you can install locally even on your laptop from `https://ollama.com`.

B. Ward, *SQL Server 2025 Unveiled*, https://doi.org/10.1007/979-8-8688-1847-9_3

The Path for AI Applications

As Muazma and Davide embarked to educate our customers on SQL and AI across 2023 and 2024 and into 2025, we came to realize that what we were talking about was often ahead of where our customers' knowledge and use of AI was. It was not because we are smarter, but it was because we had focused so much energy keeping up with the latest in AI innovations we were leaving many of the customers behind, including expert developers who use SQL.

Therefore, we produced these concepts as you see in Figure 3-1.

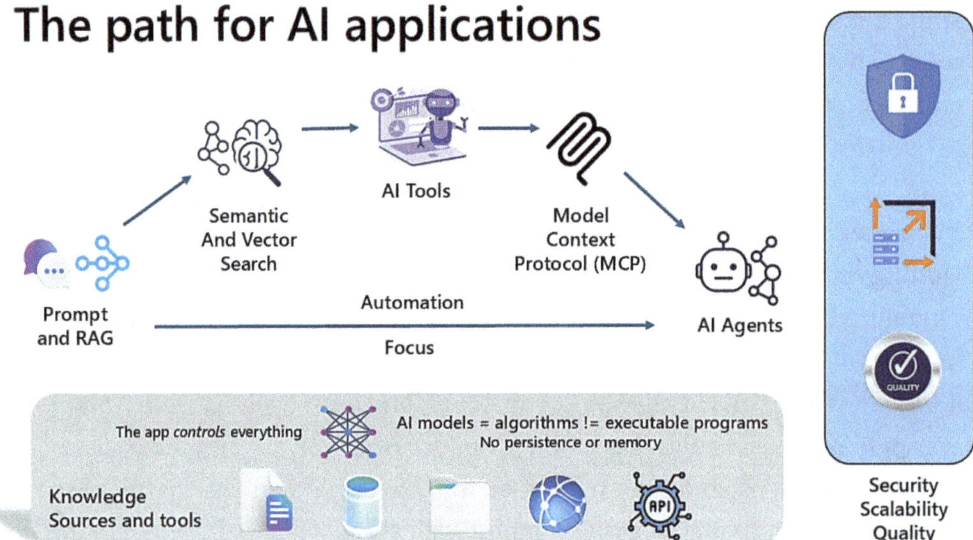

***Figure 3-1.** The path for AI applications*

The Path

Rather than have you start to the right the latest talk of AI Agents, we propose that developers can start on the left and build their way to an AI Agent. This is why we call it a path. Each piece from left to right builds on each other. Notice also the *focus* must increase as you move from left to right. The focus should increase for security, scalability, and quality. You will learn about each part of the path in the rest of the chapter.

You Control AI

One of the reasons I got very passionate about AI fundamentals was presenting at an event in 2024 when someone asked me after my talk, "How do I turn off AI?" This person's comments were not about the societal implications of current AI technology. In fact, I won't tackle this topic in this chapter or the book (although I do talk about "responsible AI" later in the chapter). This person was concerned about how AI controls software. So I asked this person for an example. They talked about how ChatGPT has a "memory" of your conversations and how they were concerned about AI arbitrarily executing code.

This is why you see in Figure 3-1 the following facts about AI models (I will describe in more detail exactly what I mean by AI models shortly in this section of the chapter):

- AI models are not executable programs. They are very sophisticated algorithms that are typically bundled in a set of files *used by applications.*
- AI models on their own have no state. This means they don't have any memory or the concept of persistence.
- AI models can't run code. They cannot execute the programs you develop. However, they can *generate code* that they can recommend you run.
- Since AI models are not executable code, they can't search the internet or look at your data. An application must perform those actions and send this information to an AI model.

What this means is that applications like ChatGPT and Microsoft Copilot are made up of executable code that uses AI models. They both offer the ability to use prompts to interact with the application. But the prompt is not interacting directly to the model itself; the application uses a prompt in many cases to send to an AI model.

Note Because ChatGPT hit the world by storm in 2022 using a Generative Pre-trained Transformer (GPT) model, many people just think ChatGPT is "AI." Many of my friends whom I know not in the technology industry believe ChatGPT itself is an AI model. ChatGPT reportedly gained 100 million users within two months. They reportedly today have 800 million active users a week.

AI models look like perhaps they have conversations, but behind the scenes AI applications are sending an entire conversation the user has with the application. This means many AI models, which are very well trained, can provide richer responses based on a series of prompts and responses from the application.

This means you as a developer have more control than you know over AI models. When I have explained these basic concepts to developers, they have a different perspective on the use of AI.

Note It is very important to know that there is an entire study of thought on how to use AI applications like ChatGPT or Microsoft Copilot responsibly. For example, you should carefully consider how these applications retain your conversations, your prompts, and even your data. It is very important to know how you as a user of AI applications can control their behavior and how they use AI models.

Knowledge Is the Key

AI models are trained on a specific set of knowledge and data that can be incredibly vast. But when an AI model is published for use, it only has knowledge of the information from that point in time it was trained. Therefore, in order to use AI models effectively, you can often augment the model's knowledge with *your* knowledge. This can be almost everything including files of any type, folders of files, database data, web search results, or even results from a program you call from an API.

Security, Scalability, Quality

Like any application that uses knowledge or data, especially private to your business, security, scalability of performance, and quality of results are critical for success. And of these three areas, security (which includes privacy) is the most critical. In fact, the biggest blocker I've seen for the adoption of AI today is security. I remember seeing this quote from our CEO Satya Nadella, "If you are faced with the tradeoff of security and another priority, the answer is clear: **Do security**."

You will see throughout this chapter and Chapter 4 how using SQL Server can help you achieve your goals for security and scalability. I'll give some thoughts about quality using AI later in this chapter.

Note What is "new" for SQL Server 2025 is built-in capabilities to support the first two aspects of the path: RAG and vector search using the data in your database. Tools, MCP, and AI Agents can all still use SQL Server 2025, including vector search, but these topics are independent of "what's new" in SQL Server 2025. Therefore, I will cover in this chapter those topics, but the focus for Chapter 4 is specifically on what is new for vector search.

Prompts and Retrieval Augmented Generation (RAG)

When you use an application like Microsoft Copilot, you interact with the application using an edit box to type in a **prompt**. Your prompts can be text in any language. When an application sends a prompt to an AI model, often called a language model (because it understands language), it does this through an API. The application uses the AI model, sends a prompt through an API, and gets a response. You can see this interaction in Figure 3-2.

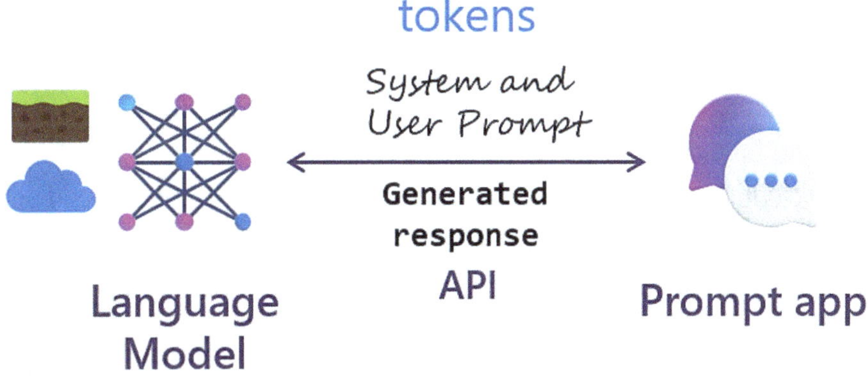

Figure 3-2. *Prompts and AI models*

Your application can either host an AI model (using known libraries) or access AI models through REST from an AI hosting service. Models can be hosted on-premises (ground) or in a cloud. This is an important security concept as you study AI models.

When you use a language model with a prompt, you typically have two prompts:

System message - This prompt helps describe how a model should behave or respond to any user prompt. For example, "You are a helpful shopping assistant."

User prompt - Your prompt for a query or directive. For example, "Help me find the best clothes for skiing."

Using an API with a model, the model can provide a **response**. Notice in Figure 3-2 the response is a *generated response*. Modern-day AI models are known as *generative* AI models. This is because these models can "generate" a response that often seems more than your prompt. In my opinion, generative AI is why AI models seem more human-like. Not only are these models trained on your language, but they are trained to respond in a way humans would.

Note I didn't want to make the visuals too complex at this point, but the API used by an application uses an **inference engine** that has the capabilities to load an AI model, accept user input, and *use* the AI model to get a response (it is a bit more complicated than that). There is no standard for inferencing engines, but a few common ones are the ONNX runtime, TensorRT, and vLLM. As a developer you can be completely abstracted from this via a hosting service that will use an inferencing runtime, but you will access AI models via REST. Examples include Ollama and cloud services like OpenAI and Azure AI Foundry. These cloud services like OpenAI and Azure AI Foundry use their own custom inferencing engines. If you want to geek out and see an example of how to build an inferencing engine, check out the Llama C++ project at `https://github.com/ggml-org/llama.cpp`. And then go download and use LM Studio (thank you, Joe Sack, for that pointer) that uses Llama C++ at `https://lmstudio.ai/`.

Any use of a model with prompts comes in the form of **tokens**. Tokens are important because specific models have limits on the amount of tokens that can be used in a prompt/response interaction. A general rule using characters for prompts is that ~100 words in English translate into 75 tokens. Token limits are the sum of the system message, user prompt, and response. These limits are documented by each model but become very important to understand if an application tries to augment a prompt with a large amount of data or attempts to send a large amount of text in a prompt as part of a conversation with a user.

To give you some context, the latest AI model for OpenAI (at the time of the writing of this book), GPT-5, supports 128,000 tokens, which is equivalent to ~192 pages of text!

To improve the responses of AI models with prompts, let's add to prompts the concept of Retrieval Augmented Generation (RAG) as seen in Figure 3-3.

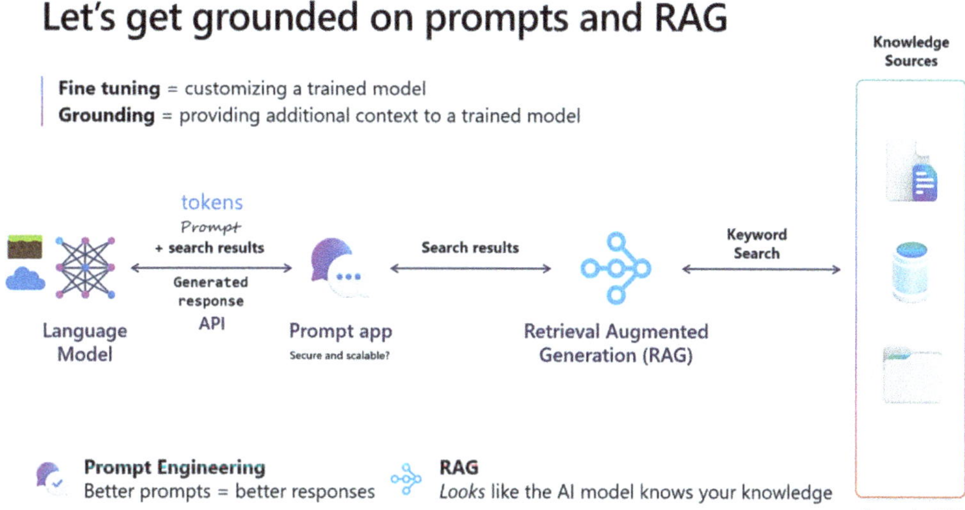

Figure 3-3. *Prompts and RAG*

RAG is a concept where you as a developer will retrieve some type of information through a search—typically the most basic form is a keyword-based search—and use these results to augment a prompt. This prompt is now richer in context to provide to the model, which can result in a richer generated response. This is also known as a *RAG pattern.*

The concept of prompt engineering is to improve prompts to provide better responses. RAG is a form of prompt engineering. Using RAG as an application can make it feel like the AI *knows* your knowledge, when in fact it was never trained on this knowledge. I believe in many cases, when an application uses a RAG pattern, the end user may be fearful that "AI" has access to their data. The facts are the application has access to the data, not the AI model. This is why the security and scalability for the application to *access* the knowledge are important for the security and scale of any AI application.

Looking back at the original prompt, you might use, "Help me find the best clothes for skiing." This prompt on its own to an AI model might seem really rich because an AI model could have been trained on the knowledge of the best clothes for skiing in general. However, if you as a developer could execute queries against your product database for ski clothes and feed this to the model along with the prompt, the model could respond with very specific recommendations based on the product's ski clothing line.

Note We will see how to make this search richer using AI models in a later section with a vector search.

The concept of using RAG is also called grounding a model. It is important to know a few facts at this point:

- Grounding is **not tuning** a model. Tuning a model is literally adding to the model's knowledge or retraining it. If you want to see an example of model tuning, check out `https://learn.microsoft.com/azure/ai-foundry/openai/tutorials/fine-tune`.
- As I've said earlier, an AI model has no persistence so does not retain your knowledge used from the RAG pattern. Only the application can retain this information.

Prompt and RAG are the first part of your path for AI applications. Let's stop and learn more about AI models before we look at the second part, vector search.

AI Models

To this point I have used the term *AI model* and even *language model*. Let's be a bit more descriptive looking at model types, sizes, and how to access models.

AI Model Types

The most popular and common AI models for applications today are the following.

Chat Completion

Chat completion models are the most popular language models in the industry. In fact, the recent "AI boom" really started with a language chat completion model called Generative Pre-trained Transformer (GPT) developed by OpenAI. GPT got is name from these characteristics (Microsoft Copilot helped me build these definitions):

- **Generative**

  The model is designed to *generate* text by predicting the next word (or token) in a sequence.

- **Pre-trained**

  It is trained on a large dataset of text before being fine-tuned (if needed) for specific tasks.

- **Transformer**

  It uses the *Transformer* architecture (see this paper: `https://arxiv.org/abs/1706.03762`) using a *deep* machine learning model (see more what deep means at `https://www.deeplearningbook.org`).

Note Bidirectional Encoder Representations from Transformers (BERT) from Google came out around the same time as GPT. BERT has similar but different capabilities than GPT (`https://arxiv.org/abs/1810.04805`).

There is no question that publicly GPT is the model that took the world by storm and has evolved and grown in size and capabilities, which I'll describe later in this section of the chapter on AI model sizes.

One unique aspect of chat completion models is the ability to have conversations or *multi-turn prompts*. Looking back at Figure 3-2 as you learned about system messages and user prompts, chat completion models support an **assistant** prompt (which is effectively the response). An assistant prompt allows an AI model to see a previous response it provided from a prompt. The next user prompt is the response from the user itself. You can provide a series of assistant and user prompts to form a chain of a conversation or *multiple turns* of prompts. This capability of chat completion models in my opinion makes them look almost humanlike. But remember the AI model has no memory, so an AI application is providing the entire history with each *turn*. This is why token limits can affect how much of a prompt conversation the application can provide to the user.

Deep Reasoning

Deep reasoning models have the same capabilities as chat completion models but far more. Deep reasoning models are designed to perform more complex "thinking" from input and can be optimized to use a concept called tools, which I'll explain later in this chapter. At the time of the writing of this book, OpenAI had launched GPT-5, which is an example of a deep reasoning model.

Note I am a user of AI applications like ChatGPT and Microsoft Copilot. Both of these applications as I was writing this book offered GPT-5 as an option. In my experience, GPT-5 provided a much richer experience than previous versions of GPT. The tradeoff was a longer time for the model to process my input. In many cases, the wait was well worth it.

Embeddings

AI embedding models have one purpose: take input in the form of text, images, audio, or video and transform them into numbers (called embeddings) that have *semantic* meaning. The numbers are called **vectors**, which are an array of floating-point numbers.

Data scientists have been using vectors, called *feature vectors*, to represent objects for various purposes. For example, a feature vector could represent the attributes of a dog. In this case the numbers are well-known by the data scientist and any human looking at the numbers.

An embedding is a vector generated by an AI model that has semantic meaning of let's say the word "dog" (or image of a data). The vectors generated by an AI model are *only known to the model*. Anyone reading these numbers would not have any idea of what they represent. Since the vectors have meaning, two sets of vectors can be compared to see if they are *similar* to each other. One common technique to perform this is by distance using a cosine math function. The closer the distance, the more similar the vectors. I'll discuss more about this concept to the next level later in the chapter with a concept called **vector search**. One of the most popular embedding models I often use is from OpenAI called **text-embedding-ada-002** (and other new models of this name).

Other Model Types

There is also wide range of other AI model types for specific purposes including but not limited to speech recognition, image generation (DALL-E is a great example), text to speech, speech to text, and language translation. That is just the tip of the iceberg. Given this chapter is about AI fundamentals, I focused on models you will often use with Microsoft SQL. But if you want to go further, check out AI model catalogs from places like Azure AI Foundry (`https://learn.microsoft.com/azure/ai-foundry/concepts/foundry-models-overview`) and Hugging Face (`https://huggingface.co/models`) or in many different GitHub repos. You will learn later in this section where the most common locations to find or use AI models.

AI Model Sizes

After the initial release of GPT and BERT, OpenAI accelerated the size and capabilities of their model with the release of GPT-2, which is unofficially the first popular **Large Language Model** (LLM) and thus begun the use of the term LLM for almost any language model. Technically speaking a Large Language Model is defined as one that is trained with *parameters* in the tens of billions and on vast datasets requiring typically significant computing resources such as GPUs.

The AI model industry has pivoted a bit, and now you can find both Medium Language Models (MLMs) and Small Language Models (SLMs). You even see models that are designed as "mini" models such as GPT-4.1 Mini and Phi-4 Mini (from Microsoft). These are "very small" models. As you move from mini to small to medium to large, you get more capabilities, but it comes with more cost and resource requirements. In my opinion, I often see everyone refer to a LLM and even use LLMs when they may not need this for their AI application.

Note Here is an in important question to consider. Do I need a GPU? In my experience, if you are fine-tuning a model, you absolutely will want a GPU. If you are using a chat completion or deep reasoning model, your performance can be significantly faster with a GPU. A lightweight use of embedding models may not require a GPU at all, but for large batches of embedding model processing, it can significantly speed up performance.

Where Do I Get AI Models and Access Them?

I mentioned earlier catalogs for AI models, and I've also talked about using an API with an inferencing engine to *use* AI models. But how do you access these models, and what does the API look like?

There are many ways to access the AI model of your choice and to access an API, but many choices today involve a hosting service where you can use a **REST** call over HTTP or HTTPS to access these AI models. Furthermore, many of the hosting services are cloud-based with a *managed* AI service, but you can also do this locally. Let's take a look at some of the most popular ways to access AI models.

Azure AI Foundry

Azure AI Foundry is a unified Azure platform-as-a-service offering for enterprise AI operations, model builders, and application development. Azure AI Foundry unifies agents, models, and tools under a single platform that is all cloud hosted. One of the capabilities for Azure AI Foundry is to allow you to deploy AI models of almost any type and access them via REST. Azure AI Foundry handles the inferencing engine and all the hosting and execution of the AI models.

All access to AI models is through the cloud service (i.e., not downloadable) and requires an Azure subscription. Azure AI Foundry used to only support OpenAI models through a curated version called **Azure OpenAI.** Since that time, Azure AI Foundry has added models from NVIDIA, Hugging Face, Mistral, Llama, Cohere, Grok, DeepSeek, and others.

Besides managing all the execution and hosting of AI models, one of the advantages of using Azure AI Foundry models is that Microsoft has a specific process to allow models to appear in the Azure AI Foundry catalog including quality, security, and fairness.

Get started with Azure AI Foundry models at `https://learn.microsoft.com/azure/ai-foundry/concepts/foundry-models-overview`.

OpenAI

As I described earlier, many consider OpenAI as one of the major pioneers of generative AI models with the lineup of GPT models. While Microsoft offers versions of OpenAI models through Azure AI Foundry, you can also purchase a subscription to access OpenAI models through their hosting service at `https://www.openai.com`. While the GPT lineup of chat completion and deep reasoning models are very popular, popular embedding models including text-embedding-ada-002, text-embedding-3-small, and text-embedding-3-large are also available. Like Azure AI Foundry, OpenAI models are accessed via REST. OpenAI has set the standard for the protocol of using REST for AI models so much that many hosting services on-premises and for cloud services support an OpenAI-*compatible* REST endpoint. Azure AI Foundry uses a version of this called Azure OpenAI REST. I'll describe later in this chapter more about how to use REST to access AI models.

Note During the writing of this book, OpenAI announced new open source models that you can download from Hugging Face or GitHub. See more at `https://openai.com/open-models/`.

Hugging Face

Starting out as a group of developers to build a friendly chatbot for teenagers (hence the friendly emoji icon), Hugging Face, `https://huggingface.co`, is the standard host of open source AI models in the world. No one knows the exact number, but it is estimated Hugging Face hosts well over 1 million AI models of every type.

Since Hugging Face models are open-sourced, many of the models allow you to download and use these models for free, hosting them in your own application or using them with a hosting service, like Ollama. Some models though have specific licensing requirements. Hugging Face also now offers a paid subscription service where you can access AI models through a web hosting service called Hugging Face Generative AI Services (HUGS).

NVIDIA

Everyone believes that NVIDIA is all about hardware and its GPUs are the backbone of AI today. But NIVIDA also has built a software platform to host and allow the consumption of AI models called the **NVIDIA AI Enterprise Platform** (`https://www.nvidia.com/en-us/data-center/products/ai-enterprise`). Developers can access this platform for free and use AI models over a hosted service via REST or download models via NVIDIA Inference Microservices (NIM), which is a container pre-built with AI-enabled models. These containers use a host service software called Triton. Triton can be independently deployed on any Linux OS and can support OpenAI compatibility via REST. NIM containers have an advantage since everything is pre-installed in a container to be deployed on Linux or Kubernetes. NVIDIA offers a catalog of AI models including chat completion, deep reasoning, and embedding models including Mistral, Llama, Phi, GPT OSS, and DeepSeek. So NVIDIA provides another option to consume AI models on-premises using REST. The NVIDIA platform does require a paid subscription to use these models outside of developer scenarios.

ONNX Runtime

Open Neural Network Exchange (ONNX) is an open source neutral format for AI models developed by Microsoft and Facebook in 2017. A program that can load and allow a developer to use an ONNX model is called an **ONNX runtime**. Any developer can build a service or program to host the ONNX runtime, which you can read about at `https://onnxruntime.ai`. While there are some AI models available in ONNX format already, for example, chat completions, there are techniques to convert an existing model to ONNX format (learn more at `https://github.com/microsoft/onnxruntime-genai/blob/main/src/python/py/models/README.md`).

A big advantage of using an ONNX runtime is you can consume the model almost anywhere, but must do so using an API hosted in your application instead of remotely using REST. This can have performance advantages for your application. Some services use ONNX behind the scenes including Foundry Local. You will see in Chapter 4 of the book that SQL Server 2025 will provide a method for you to access ONNX models hosted through the extensibility framework of SQL Server.

Foundry Local

Based on the success of Azure AI Foundry, Microsoft announced at the Build conference in May 2025 a method to use AI models on Windows *and* macOS devices called Foundry Local. The concept is to bring the world of Azure AI Foundry for model hosting and consumption to Windows and macOS devices.

Foundry Local is free to download and includes a catalog of both free, open source AI models and some models that require a license to download and use. A huge benefit of Foundry Local is that models are optimized for the hardware it can detect including CPUs, GPUs, and NPUs.

Foundry Local includes an SDK to host models locally or access via an OpenAI-compatible REST endpoint. What is interesting is behind the scenes Foundry Local uses the ONNX runtime so all models that work with Foundry Local have been converted to ONNX. This mean you can "bring you own model," but it must be in an ONNX format.

Learn more about Foundry Local at `https://learn.microsoft.com/azure/ai-foundry/foundry-local`.

Ollama

As I write this book, the most popular AI model hosting software in the world for on-premises is **Ollama** (`https://ollama.com`). Ollama was created in 2021 to promote a secure method to host and consume AI models as an open source project. Ollama runs on Windows, Linux, and as a container. Ollama runs as a service, and all access is through REST (either directly or through SDKs). This means Ollama can be accessed on the same computer as your application or over a network (and technically in a cloud-based virtual machine (VM)). You can download a default set of AI models with Ollama (see the current list at `https://github.com/ollama/ollama`), which you can easily configure to be optimized for GPUs. You can also bring your own model provided it is in GPT-Generated Unified Format (GGUF) (`https://github.com/ggml-org/ggml/blob/master/docs/gguf.md`). You can download models from Hugging Face that support this format or convert it yourself (see an example at `https://www.geeksforgeeks.org/machine-learning/how-to-convert-any-huggingface-model-to-gguf-file-format`). Ollama has now ventured into a paid subscription service to use faster, larger models at `https://ollama.com/turbo`.

Tip When I install Ollama on Windows, it comes with a client program that you can use a chat test application. It is a great way to see in action with a user interface how to test various AI models.

I'll use Ollama in the next section of this chapter and in Chapter 4 of the book with SQL Server 2025. Download and install it at `https://ollama.com/download`.

Let's REST with AI

The Representational State Transfer (REST) protocol became widely adopted in the mid-2000s for web applications using HTTP. REST is also a great way to communicate to any service including AI hosting models through an *endpoint*, which is a URL (whether it is local, network based, or across the web). There are many different SDKs and libraries to use REST for just about any programming language. One of the nice independent methods to test any REST service is the program **curl** (`https://curl.se/docs/manpage.html`). Curl is so popular it just comes by default with almost any OS installation including Windows. There are several examples in this section of the chapter that you may want to try out and follow along. It could be very helpful to go through these to make it easier to follow examples with SQL Server 2025 and AI in Chapter 4.

Prerequires for the Examples

I created these examples on Windows, but you can also do this on Linux.

1. Download and install Ollama from `https://ollama.com/download/windows`.

2. Download a proxy software like **caddy** from `https://caddyserver.com/download`. You will need this for HTTPS testing. I prefer caddy to use locally over nginx because self-signed certificates are automatically installed.

    a. Create a text file called **Caddyfile**. This is a configuration file for caddy including directives to redirect traffic to Ollama. The file should contain this text:

```
{
  admin localhost:2019
}
localhost, 127.0.0.1 {
  tls internal
  @ollama path /api/* /v1/*
  reverse_proxy 127.0.0.1:11434 {
    flush_interval -1
  }
  @preflight {
    method OPTIONS
    path /api/* /v1/*
  }
  respond @preflight 204
  header @ollama {
    Access-Control-Allow-Origin  *
    Access-Control-Allow-Methods "GET, POST, OPTIONS"
    Access-Control-Allow-Headers "Content-Type, Authorization"
    Access-Control-Expose-Headers "*"
  }
  respond / "Caddy is running with HTTPS on localhost" 200
  log {
    output file C:\caddy\logs\ollama-local.log
    format console
  }
}
```

b. Start caddy with the following command where you installed it:

```
.\caddy_windows_amd64.exe run --config Caddyfile
```

c. Now in another command window, set up caddy so it's installed certificates are trusted on the computer or VM (sometimes I've not needed this step):

```
.\caddy_windows_amd64.exe trust
```

REST with AI Using Ollama

Let's use curl test out a few AI model scenarios with Ollama. After you have installed Ollama (I'll use Windows for my examples), run the following steps to access **llama3**, a chat completion model, via REST. Ollama is running already to accept requests for a model when you install it.

1. Preload the llama3 model:

   ```
   ollama pull llama3
   ```

2. Use curl from the Windows Command shell as follows:

   ```
   curl -H "Content-Type: application/json" -d "{\"model\":
   \"llama3\",\"messages\":[{\"role\":\"system\",\"content
   \":\"You are a helpful assistant that explains database
   concepts clearly.\"},{\"role\":\"user\",\"content\":\"Ex
   plain the difference between clustered and nonclustered
   indexes in SQL Server.\"}],\"stream\":false}" http://
   localhost:11434/api/chat
   ```

   The -H parameter are the "headers" and the -d is the "data" for a POST command. In this case the data is a JSON-formatted text with the name of the model to use and messages, which include a system message and user message (denoted by "role"). The "stream:false" tells Ollama to provide all response all at once.

   My laptop has a GPU, so the response came back quickly (if you don't have a GPU, this response could take minutes). When the response comes back, it looks like this (I've removed the entire response and just included part of it):

   {"model":"llama3","created_at":"2025-09-02T14:31:36.6700267Z" ,"message":{"role":"assistant","content":"In SQL Server, an index is a data structure that improves query performance by speeding up the retrieval of specific rows from a table or view. There are two main types of indexes: clustered and non-clustered.... <text omitted here>...Remember that choosing the right type of index depends on your specific database schema, query patterns, and performance requirements."},"done_reason":"stop","done":true," total_duration":83009741300,"load_duration":55966500,"prompt_eval_count":41,"prompt_eval_duration":129398700,"eval_count":468,"eval_duration":82822261500}

You can see in the response there is metadata at the front and end of the overall response. The heart of the response is the form of a message where the role is **assistant**. The "content" is the actual response from the AI model. This is a great example of a "raw" conversation with a model. In this case Ollama is accepting messages from its local URL endpoint, `http://localhost:11434/api/chat`, and using its inferencing engine code to interact with the llama3 model. The **/api/chat** is an Ollama protocol for using REST (see the reference at `https://deepwiki.com/ollama/ollama-js/3-api-reference`). Azure OpenAI and OpenAI (including OpenAI-compatible endpoints) use **/v1/chat/completions**. The OpenAI documentation for chat completion REST protocol can be found at `https://platform.openai.com/docs/api-reference/chat`. The Azure OpenAI reference is at `https://learn.microsoft.com/azure/ai-foundry/openai/reference`.

You can now see the way to use REST with chat completion models. You will see in Chapter 5 of the book how to harness this knowledge to use the T-SQL procedure **sp_invoke_external_rest_endpoint** to send REST to a chat completion model.

Secure REST with HTTPS

In the examples above the endpoint for Ollama used http. HTTP by default is unencrypted, which is why a secure HTTP was invented called HTTPS. HTTPS encrypts traffic between the client and endpoint, but the service supporting the endpoint must support HTTPS. Cloud services like Azure OpenAI and OpenAI support HTTPS, but services like Ollama and NVIDIA Triton do not.

This means you need to use a proxy service that will accept HTTPS and redirect the traffic to a local host service like Ollama. While nginx is a popular proxy program, I like to use caddy because it installs self-signed certificates so it makes it easy to use HTTPS. I explained earlier in the "Prerequisites for the Examples" section how to install and configure caddy. Once you have this installed and running, you can change the example above to send a prompt to llama3 with Ollama to this:

```
curl --ssl-no-revoke -H "Content-Type: application/json" -d "{\"model\":\"l
lama3\",\"messages\":[{\"role\":\"system\",\"content\":\"You are a helpful
assistant that explains database concepts clearly.\"},{\"role\":\"user\"
,\"content\":\"Explain the difference between clustered and nonclustered
indexes in SQL Server.\"}],\"stream\":false}" https://localhost/api/chat
```

In this example, notice the endpoint uses https and no longer specifies the Ollama port because caddy is listening on the default HTTPS port 443 and redirecting this to Ollama per the Caddyfile configuration. The option --ssl-no-revoke is only needed if you use the built-in curl.exe that comes with Windows. You will not need this when using T-SQL to use a proxy.

Add Some Vector Search

In Figure 3-1, the next step for the path for AI applications is **semantic or vector search**.

Vector Search and Your Data

I've discussed the concept of embeddings and AI models that support generating embeddings. Figure 3-4 shows a visual of how you could add embeddings and vector search in AI models using a database or index.

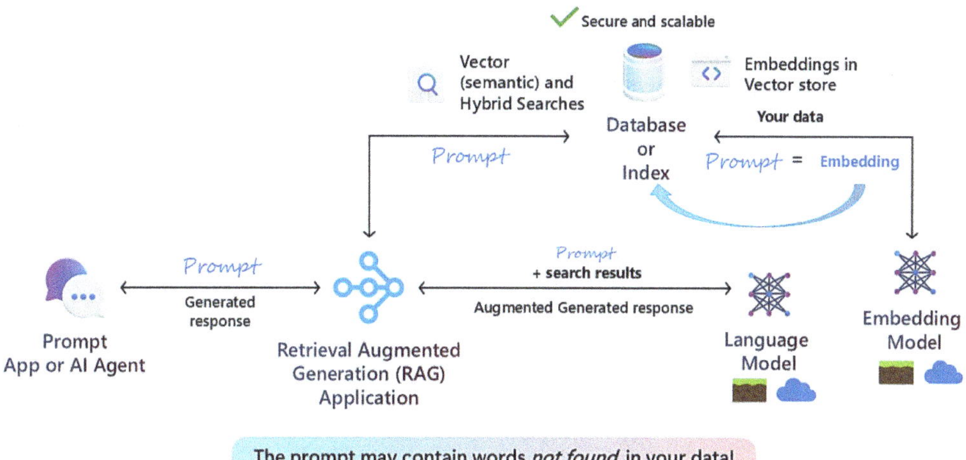

Figure 3-4. *Vector search with your data*

In this visual you can see we've added the concept of generating embeddings from an embedding model. The concept is to take text, generate embeddings, and store these in a database or index. Then you can generate an embedding for a prompt. Now you can "search" for the closest embeddings in your database or index to your prompt. I mentioned earlier in the chapter that a common technique is to use a cosine distance function (you will see in Chapter 4, SQL Server has this all built-in to T-SQL). This concept is called a *vector search,* also known as a *semantic search.* Notice the important aspect about vector searching: **the prompt may contain words not in your data.** That is the power of using embedding models since these models are trained to find out things that are similar in meaning.

The RAG pattern becomes even more effective since the results are using a vector search (or hybrid search where you can combine other search techniques like keyword searching) and being sent to a language or chat completions model.

One simple way outside of SQL Server to use vector search is with **Azure AI Search.** You can easily build a vector index on your knowledge and use any code or tools to execute a vector search. Azure AI Search has built-in support for indexers for Azure SQL Database, Azure SQL Managed Instance, and SQL Server on Azure Virtual Machines (and some third-party partners support on-premises SQL Server). You can learn more at `https://learn.microsoft.com/azure/search/search-how-to-index-sql-database`.

One simple technique I use to prototype vector searching outside of the new SQL Server 2025 vector capabilities is to use built-in support for Azure AI Foundry to load an Excel spreadsheet to "chat with my data." Foundry will create a vector index and let you start searching. Figure 3-5 shows an example.

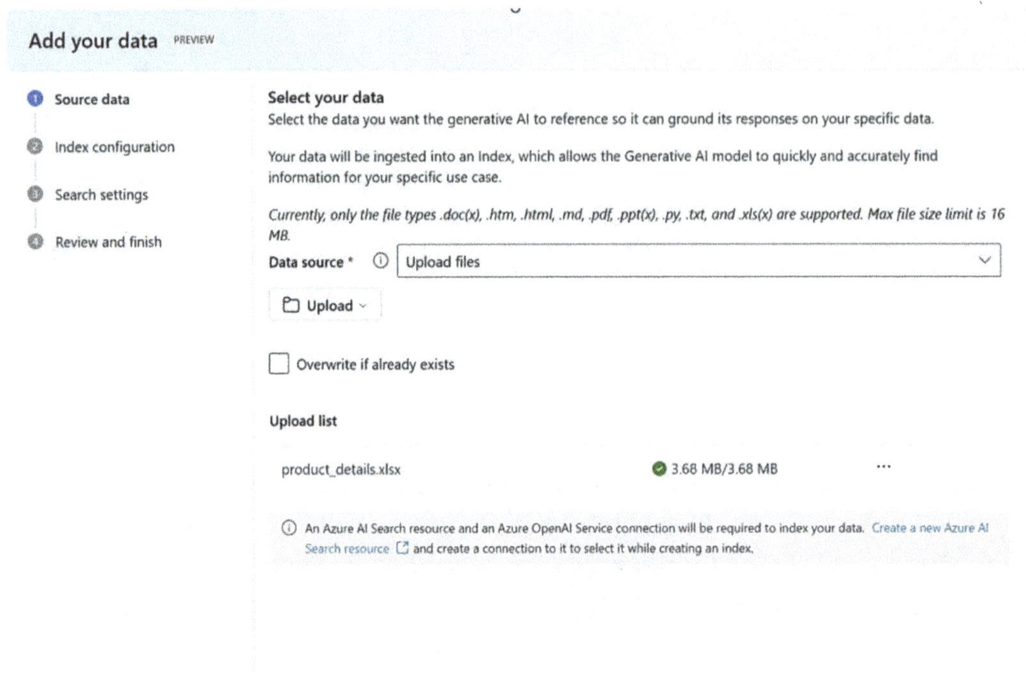

***Figure 3-5.** Building a vector index on a spreadsheet with Azure AI Foundry*

With the index in place, I can use a chat playground concept in Azure AI Foundry to use prompts that will use a vector search and RAG pattern like in Figure 3-6.

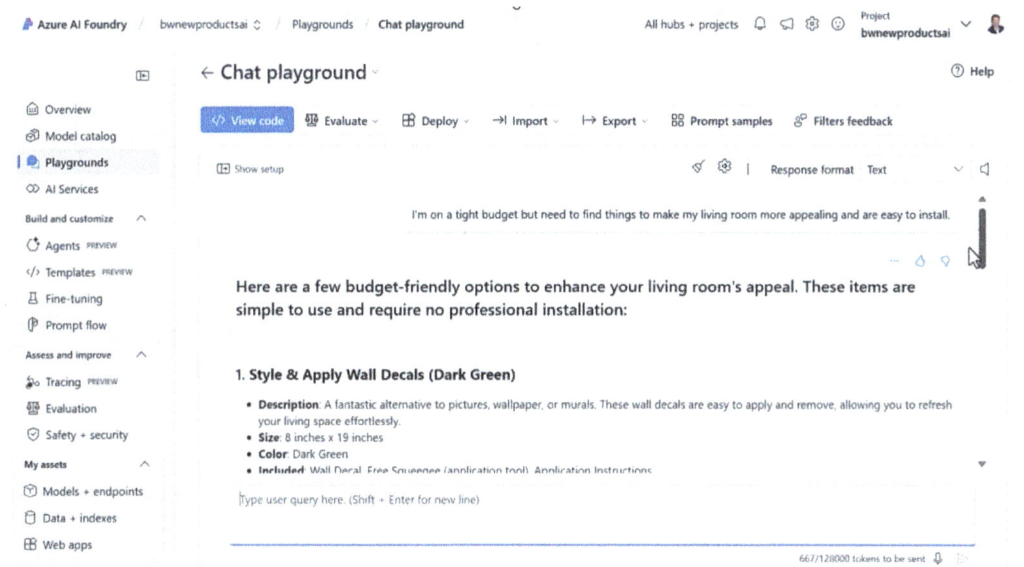

Figure 3-6. *Using the chat playground for a vector search on a spreadsheet*

Notice my prompt uses words like "tight" and "appealing," for which an embedding model knows how to generate numbers that match closer with specific products in the spreadsheet powered by a vector index.

Note You will learn more about how SQL Server 2025 supports the concept of a vector index in Chapter 4.

Let's see how REST works with embedding models to set you up for success in Chapter 4.

Use REST with Ollama to Do a Vector Search

With Ollama and caddy all set up, you can easily generate embeddings with an embedding model.

1. Pull down an embedding model from the Ollama catalog with this command:

```
ollama pull mxbai-embed-large
```

2. Use curl to generate embeddings with the following command:

```
curl --ssl-no-revoke -X POST https://localhost/api/
embeddings -H "Content-Type: application/json" -d
"{\"model\": \"mxbai-embed-large\", \"prompt\": \"The
Dallas Cowboys are the best team in the NFL\"}"
```

The response (I have not included all the results for brevity) looks like this:

```
\"mxbai-embed-large\", \"prompt\": \"The Dallas Cowboys are the best team
in the NFL\"}"
{"embedding":[0.12590645253658295,0.5879506468772888,-0.3145020604133606,
1.0422884225845337,-0.09395606815814972,-0.4105185568332672,
-0.16694705188274384,-0.5184409022331238,0.8630852699279785, ….<results
trimmed forbrevity>….-0.22674836218357086,-0.42288142442703247,
0.07240116596221924]}
```

You can see the results contain a JSON value for embedding followed by an array of floating-point numbers. These are the embedding values that represent the meaning of the text submitted. You will see in Chapter 4 how SQL Server will convert these numbers into a binary format in a new vector data type and allow you to perform vector searches inside your database!

AI Tools and Model Context Protocol

If you have built an application using prompts, RAG, and vector search, you have done something few are accomplishing today: a successful AI application. I personally believe there are so many opportunities with this AI application pattern today, and that is why what we are proposing is a "path."

AI Tools

Let's say you have conquered a successful vector search and RAG application. What's next? While vector searching is great, what about scenarios for SQL Server where you need more than just a "search"? You need queries to get precise information based on T-SQL SELECT statement options. What might be interesting is to use the built-in knowledge of the T-SQL language from an AI model to help construct T-SQL based on a user prompt. So for a user it looks like they are "chatting with their data" and can submit prompts that require precise lookups in the database. Most popular AI chat completion models are trained on T-SQL, but they are not trained *on your specific database*. One way to have a model provide information like it "knows your data" is with the RAG pattern we have described earlier. However, wouldn't it be nice if the AI model could *assist* you in building the SQL query to use for the RAG pattern? Now SQL Server becomes a **tool** for the AI model.

Figure 3-7 shows an example of how an AI application can use SQL Server and an AI model together with SQL Server serving as the tool. Notice the AI application executes any T-SQL, *not* the AI model, keeping the theme at the beginning of this chapter that the application is in control.

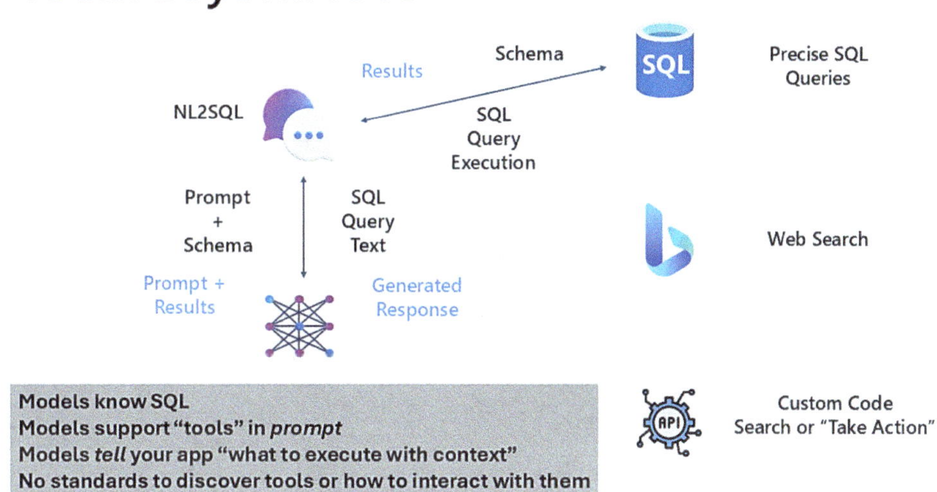

Figure 3-7. *SQL Server as an AI tool*

In this visual the application retrieves the schema needed for any SQL queries (or prompts could "hardcode" in the schema) and sends a prompt with this schema (typically in the form of a system message) to the AI model. The AI model will respond with a suggested T-SQL query that the application can decide to execute against the database. Optionally, the application could take the results of the query and augment the original prompt with the results completing the circle of a RAG pattern. The information provided to the AI model can even make suggestions about using vector search as part of generating the query if it makes sense in the context of the user prompt.

Azure SQL Database and SQL Database in Fabric are now supported with Microsoft Copilot Studio as *knowledge sources*. When you use these sources, Copilot Studio uses AI models with SQL as a tool to generate SQL statements to support user prompts.

Let's look at an example using the deployment of Ollama and caddy you have from previous examples in the chapter:

1. In order to support the tools concept, you need to download a new llama3 version called llama3.1 with this command:

   ```
   ollama pull llama3.1
   ```

2. Create a JSON file, and call it **payload.json** with the following text:

   ```
   {
     "model": "llama3.1",
     "stream": false,
     "tools": [
       {
         "type": "function",
         "function": {
           "name": "propose_sql",
           "description": "Propose a read-only, parameterized T-SQL
           SELECT for Azure SQL / SQL Server. Never perform DML
           or DDL.",
           "parameters": {
             "type": "object",
             "properties": {
               "dialect": { "type": "string", "enum": ["tsql"] },
               "sql": {
                 "type": "string",
                 "description": "One T-SQL SELECT only. Use the given
                 schema. If a timeframe is requested or implied
                 (e.g., 'last N days/weeks/months', 'recent'),
                 include WHERE with @start_date and @end_date. Use
                 explicit JOINs, list columns (no SELECT *), GROUP
                 BY when aggregating, and ORDER BY when returning
                 TOP rows.",
                 "minLength": 80,
                 "pattern": "(?is)^\\s*select\\b.*\\bfrom\\b.*$"
               },
               "bindings": {
   ```

```
          "type": "object",
          "description": "Named parameters the app will
          bind with strong types. Do not inline any values
          in SQL.",
          "additionalProperties": {
            "type": "object",
            "properties": {
              "type": { "type": "string", "enum": ["string",
              "int", "float", "bool", "datetime"] },
              "example": {}
            },
            "required": ["type"]
          }
        },
        "tables_used": {
          "type": "array",
          "items": { "type": "string" },
          "description": "List tables actually referenced in
          the SQL."
        },
        "reasoning": { "type": "string" }
      },
      "required": ["dialect", "sql", "bindings"]
    }
  }
}
],
"messages": [
  {
    "role": "system",
    "content": "You are a T-SQL generator for Microsoft SQL
    Server (Azure SQL). For every user prompt, respond ONLY
    by calling the function tool 'propose_sql' with a single,
    parameterized, read-only T-SQL SELECT based strictly on
    the provided schema. Do not output prose, code fences, or
    comments."
```

```
    },
    {
      "role": "system",
      "content": "T-SQL policy:\n- Read-only: SELECT statements
      only. Never produce INSERT/UPDATE/DELETE/MERGE/EXEC/
      DDL.\n- Parameterize all user-controlled values via named
      parameters in 'bindings' (e.g., @start_date, @end_date, @
      limit). Do not inline literals.\n- Timeframes: If the prompt
      requests or implies a time window (e.g., 'last 30 days',
      'recent'), include WHERE with h.OrderDate >= @start_date
      AND h.OrderDate < @end_date. Provide datetime2 examples in
      UTC via bindings; do not use INTERVAL syntax.\n- Top N: Use
      TOP (@limit) with an appropriate ORDER BY when the prompt
      asks for ranked or top results.\n- Joins & columns: Use
      explicit JOINs and short aliases. Avoid SELECT *; list only
      needed columns. Use GROUP BY when aggregating.\n- Dialect:
      SQL Server T-SQL (use TOP, GETUTCDATE()/SYSUTCDATETIME(),
      DATEADD in examples; no LIMIT)."
    },
    {
      "role": "system",
      "content": "Schema subset (AdventureWorks):\n- Sales.
      SalesOrderHeader( SalesOrderID int, OrderDate datetime2,
      CustomerID int, SubTotal money, TaxAmt money, Freight money,
      TotalDue money )\n- Sales.SalesOrderDetail( SalesOrderID int,
      ProductID int, OrderQty int, UnitPrice money, UnitPriceDiscount
      money, LineTotal numeric(38,6) )\n- Production.Product(
      ProductID int, Name nvarchar(200), ProductNumber nvarchar(50),
      ProductSubcategoryID int )\nNotes:\n- Revenue typically
      aggregates Sales.SalesOrderDetail.LineTotal.\n- Join
      SalesOrderDetail -> SalesOrderHeader on SalesOrderID.\n- Join
      Product via ProductID."
    },
    {
      "role": "user",
```

```
      "content": "Show the top 10 products by total revenue in the
      last 30 days."
    }
  ]
}
```

There is a lot to unpack here, but pay most attention to the following:

- A section called "tools" where now the AI model is given special instructions about what kind of tool it can use to help provide a response, which in this case is to generate a T-SQL query (not execute it).
- The various system messages (yes, you can use more than one) to give explicit instructions on what to do and what *not* to do when building a T-SQL statement to respond to a prompt and the schema of the database to use as guidance when building the T-SQL statement.
- The user prompt is like what a user might "chat" with an AI application to try and look at data without knowing how to build a T-SQL statement to do it.

3. Use the following curl command using a Windows prompt to interact with the llama3.1 using Ollama (and caddy for HTTPS):

```
curl --ssl-no-revoke https://localhost/api/chat -H "Content-Type:
application/json" --data-binary "@payload.json"
```

4. Look at your results, which should look similar to this:


```
{"model":"llama3.1","created_at":"2025-09-03T01:03:00.0603635Z"
,"message":{"role":"assistant","content":"","tool_calls":[{"fun
ction":{"name":"propose_sql","arguments":{"bindings":{"limit":1
0,"start_date":"2023-12-01T00:00:00"},"dialect":"tsql","reasoni
ng":"","sql":"SELECT TOP (@limit) h.SalesOrderID, d.ProductID,
SUM(d.LineTotal) AS Revenue\nFROM Sales.SalesOrderDetail d\nINNER
JOIN Sales.SalesOrderHeader h ON d.SalesOrderID = h.SalesOrderID\
nINNER JOIN Production.Product p ON d.ProductID = p.ProductID\
```

nWHERE h.OrderDate \u003e= @start_date AND h.OrderDate \u003c GETUTCDATE()\nGROUP BY d.ProductID, h.SalesOrderID\nORDER BY Revenue DESC","tables_used":["Sales.SalesOrderDetail","Sales.SalesOrderHeader","Production.Product"]}}}]},"done_reason":"stop","done":true,"total_duration":5612022400,"load_duration":1828518900,"prompt_eval_count":756,"prompt_eval_duration":282571400,"eval_count":181,"eval_duration":3497271200}

The "heart" of the response is the "sql" that is a SELECT statement using parameter names with "bindings" listed above. Any application could take this information, deconstruct the JSON, and use this to execute a parameterized SQL statement against the database.

You can use a combination of the "tools" payload and "system messages" to provide very prescriptive guidance for how an AI model should generate the right SQL query for the user prompt. An AI model can use its knowledge of T-SQL combined with these messages to generate the best query possible based on the prompt.

Notice in Figure 3-7, visuals for other "tools" such as web searching or any "API" calls to custom code. Tools are a general concept, so they can be formed to anything you need. The key is that the AI model can use information it is already trained on plus additional context to provide a response for the application to perform web search or call APIs with very specific instructions to make the AI application have the ability to dynamically execute code based on user prompts.

Take a second and think about the security concept you have learned so far. In these examples with tools, it is up to the AI application to use the right security models. You can even provide guidance through tools and system messages for AI models to only generate specific responses in line with security requirements of the application.

Model Context Protocol (MCP)

In the previous example, while "tools" is a common concept for the latest AI models to understand, there is no standard to discover what tools can do or protocols for applications to use these tools. This is why the Model Context Protocol (MCP), https://modelcontextprotocol.io/docs/getting-started/intro, was developed by Anthropic in November of 2024.

Often called the "USB-C for AI," MCP (which is open source) is starting to take hold as a standard to build AI applications, making it easier to build tools to use with AI models. MCP defines these three software components.

MCP Server

A software program that implements tools to be used by MCP hosts in coordination with AI models.

MCP Client

The "provider" that handles the protocol and communication between MCP hosts and MCP Servers.

MCP Host

The AI application that hosts the MCP client and uses MCP Servers to coordinate with AI models for various tools.

All communication between MCP clients and servers is through a JSON+RPC protocol over stdio (local) or HTTPS (remote).

Let's take the example earlier with tools and AI models for T-SQL queries. Microsoft has built a preview of a new MCP Server for SQL (`https://github.com/Azure-Samples/SQL-AI-samples/tree/main/MssqlMcp`).

The MCP Server provides "tools" for various tasks that are common to interact with a SQL Server or Azure SQL Database such as **List Tables** or **Read Data**. Now if an AI application wanted to accept prompts that it wants to generate T-SQL for, it could use the MCP Server as a "tool" with AI models. In this scenario, your AI application is the MCP host using the MCP client to communicate to the MCP Server. The MCP host will use results of communication from the client and server to interact with AI models. The key here is that the *MCP Server executes the T-SQL code*, not the MCP host. It provides another extra security layer for AI applications. In addition, the MCP Server can "bake in" all the best practices and rules for how T-SQL queries should be implemented by AI models.

You can build your own MCP host or try it out with various platforms that include a host and client including Visual Studio Code with GitHub Copilot and Claude desktop.

MCP is still new but catching fire pretty quickly. MCP Servers are all still open source, so this could delay the adoption of these in enterprise AI applications. The concept is sound and can create much more stable and standard methods for AI tools.

AI Agents

The last component of our path to AI applications are **AI Agents**. My colleague Muazma Zahid read that Microsoft CTO Kevin Scott defined AI Agents as "Agent is a thing a human can delegate tasks to."

> I did an interesting test. I asked ChatGPT what an AI Agent was. Here was the response:
>
> **A normal chat model for ChatGPT is something like the following:**
> You: *"Summarize my meeting notes and draft an email."*Me: I might summarize the notes if you paste them in and then I will propose an email for you to use.
> **But Agent mode is like the following:**
> You: *"Summarize my meeting notes and draft an email."*Me (in Agent mode):

1. Locate your meeting notes in Teams.
2. Summarize them.
3. Draft an email.
4. (Optionally) Propose sending it in Outlook.

Do you see the difference? AI Agents are not just about responding to prompts but to complete tasks and in some cases complete a task autonomously. But don't forget even in this example, the AI model is not doing a task. It is an AI application interacting with an AI model for assistance to complete a task.

This is why in Figure 3-1 I showed you that focus for security, scalability, and quality must increase as you move toward AI Agents.

Note Microsoft contributed to this interesting paper on information flow control with AI Agents that is worth reading at `https://arxiv.org/abs/2505.23643`.

The concept of AI Agents also promotes the concept of *modularity* of design. If you are building AI applications to perform certain tasks or functions, it can often be better to break up the problem than have one monolithic application.

My colleague Davide Mauri built a very nice example to show this in action involving an insurance company (see `https://github.com/Azure-Samples/azure-sql-db-chat-sk/tree/insurance-chatbot-demo`). Figure 3-8 shows a graphic of how AI Agents could work in coordination.

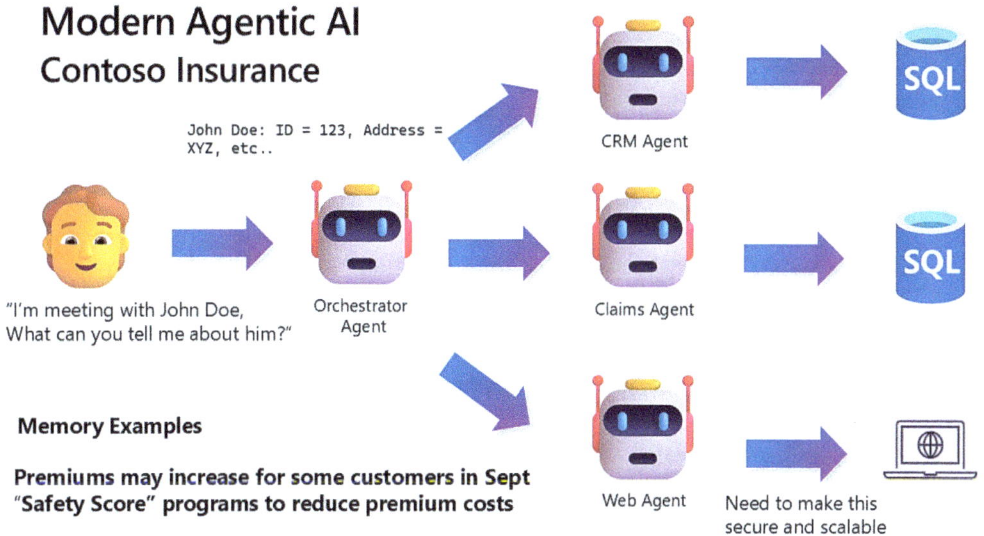

Figure 3-8. *AI Agents*

In this example, an Orchestrator agent uses other agents for specific tasks such as CRM, claims, or web searches. This allows developers to build reusable and specifically focused AI applications that can be more easily secured and scaled. Optionally, each of these agents could use MCP Servers to accomplish their tasks. Notice on the visual the concept of *memory*, which is a practice many AI Agent applications use.

Microsoft is making a big bet on the future of AI Agents. Learn more at `https://enablement.microsoft.com/ai-agents/`.

What About Quality?

In Figure 3-1 I mentioned quality. Here are a few considerations when it comes to quality, AI applications, and your data:

- **Look at AI model benchmarks.**

  There are benchmarks for AI models. Consider keeping up with this site to see the "leaderboard" for AI models: `https://llm-stats.com`. In addition, Azure AI Foundry has excellent resources available in the Azure Portal to compare models for quality and performance. See more at `https://learn.microsoft.com/azure/ai-foundry/how-to/benchmark-model-in-catalog`.

- **You control the AI model and which one to use.**

  You read in this chapter how to access and use models. The decision is up to you. Even AI applications that you use in everyday life now can give you choices on certain models to use. As part of writing this book, I used Microsoft Copilot with the option of "Try GPT-5" instead of GPT-4.1.

- **Research the training behind the models.**

  All models are trained on some base of knowledge. But not all models are trained on the same knowledge or the same way. I would personally research any details on how a model was trained before using it in production. For example, here is a great resource on the training behind the GPT-5 system: `https://cdn.openai.com/gpt-5-system-card.pdf`.

- **Your use of AI models can be tuned for quality, accuracy, and consistency.**

  Your use of AI models can be tuned (without actually fine-tuning the model) based on the parameters you send the model. I've shown you examples of using system messages and prompts, but most models support other parameters such as **temperature**. Temperature controls the randomness or creativity of the model's responses. Numbers typically range from 0 to 2.0 with 1.0 being the default. The lower the temperature, the more consistent the model responses to the same prompt multiple times. The higher the number, the more creative the model responses with the cost of possible inconsistency or even what may seem like *hallucinations*.

- **You control the quality of your application, data, and tools.**

  Like any software system, you ultimately control the quality of your application, your data, and the tools you use. Don't let the fact you are using an AI model now as part of the equation change what you are already doing today.

What Have We Learned

I thought it would be helpful to include some lessons our team at Microsoft has learned working with our customers:

- **Create a responsible AI policy.**

  Each time I present on the topic of AI, I ask my audience who among them works for a company that has a **responsible AI policy**. I'm amazed at how few people have one. Microsoft has a clear one, which you can read at `https://www.microsoft.com/ai/responsible-ai`. This includes a detailed transparency report at `https://cdn-dynmedia-1.microsoft.com/is/content/microsoftcorp/microsoft/msc/documents/presentations/CSR/Responsible-AI-Transparency-Report-2025-vertical.pdf`. If you don't have a policy, consider starting now. It will transform how your company thinks, uses, and accesses AI.

- **Treat AI models like software.**

  I've said earlier in this chapter AI models are not executable software programs. However, you should treat models the same way you treat any software you purchase or consume today. In many large enterprises there are precise guidelines on software procurement. If you are not including AI models in that process, start building a plan now. This could include conversations about using an AI model that is open source.

- **Spend time designing the right solution with guardrails.**

  We have seen so many customers aggressively work on AI solutions without really knowing the problem they are trying to solve. I've heard "We just need to use AI" or "We need a chatbot agent" without understanding the underlying business problem. Because of this I've seen many AI projects fail, which may deter the company from pursuing the right solution in the future. Consider the path I've described in this chapter as a way to start small and grow with more advanced AI technology as you see success. Apply the same rigor (with room for prototyping and proof of concepts) as you do with all software projects.

- **Choose the right AI models for the right solution.**

  I've seen customers get frustrated when they are trying to use a chat completion model to generate embeddings or vice versa. Learn more about model types and understand which one you need to build a particular kind of solution for your business.

- **Review your data quality and security.**

  The quality and security of your usage of AI with your data will only be as good as the quality and security of your data. This is the same with any type of new solution you are trying to build.

Launch SQL Server 2025 AI Built-In

Knowledge is the fuel for AI—the knowledge you bring as you learn about AI and the knowledge sources you use with AI models. This chapter is intended to provide you fundamental and foundational knowledge to empower you to learn the details of how SQL Server 2025 provides built-in AI capabilities with vector search including a very unique way to allow you to access AI models securely and use the power of the SQL Server engine for scale.

I asked Muazma Zahid her perspective on the importance of AI for the future of any SQL professional: "*The AI age is here. SQL professionals must evolve beyond traditional query tuning and schema design. Understanding AI fundamentals is no longer optional, it's essential. SQL Server 2025 integrates AI deeply into the engine, enabling developers and DBAs to build intelligent applications, automate insights, and unlock new dimensions of analytics. Learning AI empowers you to harness these capabilities and future-proof your career.*"

CHAPTER 4

AI Built-In

The SQL Server team was thinking about *AI built-in* before AI was a "thing." In SQL Server 2016 we introduced **SQL Server 2016 R Services**. This feature allowed developers to run an R program in a *sidecar* fashion with the SQL Server engine. When you installed this feature, we deployed a separate set of services on Windows to allow an R program to be integrated with data in SQL Server and execute code to run a machine learning model. Internally we called this feature **SPEES**, which stands for **sp_execute_external_script**. This is a system procedure you used to run your R code you installed locally on the VM or computer where SQL is installed. Our thinking was to bring the world of data science to SQL Server instead of running on client computers. SQL Server could be used to securely run the R code in a very scalable way. We even supported resource governor workgroups to control resource usage.

These services used what we call an *extensibility framework* to run isolated from SQL Server but well integrated to exchange data inside the database. In SQL Server 2017, we renamed this feature to **Machine Learning Services** adding support for Python and Linux. In 2019, we extended the architecture to support Java, which we called **Java Language Extensions**. In all of these cases, the developer was required to train a model to use for scenarios like predictions, ranking, and forecasting. In 2017, we also introduced a T-SQL **PREDICT** function to load and run pre-trained ONNX and R models. We made a big splash about these features, and we thought this was the future of using SQL with AI. While these features were used, they never really took off as we thought they would.

Now fast forward to 2022 and 2023, where pre-trained generative models now supported humanlike interaction vs. having to write R or Python code to access machine learning technology. And of course, these models are so much more powerful. All of this work did not go to waste as it allowed our team to be more familiar with AI concepts and even lay the groundwork to reuse the extensibility architecture, which you will see in this chapter.

B. Ward, *SQL Server 2025 Unveiled*, https://doi.org/10.1007/979-8-8688-1847-9_4

In this chapter you will learn end to end how SQL Server 2025 provides AI built-in with enterprise-grade security and scale through support of **vectors**. You will learn what types of problems we are trying to solve, what features we are providing to solve them, how the vector architecture works, and examples of using AI models ground to cloud. By the end of this chapter, my goal is for you to see why SQL Server 2025 is the **AI-ready enterprise database**.

There are various T-SQL examples in the book. Some of them are "inline" to show a feature. Some are "full examples" with scripts. All scripts for these can be found in the **ch4 - AI** folder. Besides needing SQL Server 2025, I will list all the perquisites in the sections of the chapter for the examples.

What Are We Trying to Solve

When I first started presenting our concepts of vectors with SQL Server 2025, I quickly realized I was presenting the solution without talking about the problem. It is our intention with SQL Server 2025 to solve these types of problems.

Smarter Searching

SQL Server supports searching of text data through LIKE clauses, other T-SQL functions, and full-text search (FTS). If you read Chapter 3, you saw that a vector search can provide a new way to use a semantic search on your data including finding data using words or phrases *not in your data*. So if you have databases with existing text data, vector search can provide richer and more robust searching capabilities.

Support Centralized Vector Searching

We showed this to several customers, and they started thinking in terms of centralizing vector searching with text in SQL Server 2025. They started looking into taking existing documents and text and bringing them into SQL Server instead of doing vector searches across multiple data sources. These customers liked the security model of SQL Server and love the T-SQL language. My colleague Davide Mauri built in an example pipeline to take PDF and Microsoft Word documents to bring them into SQL Server to support vector search. You can learn more at `https://github.com/Azure/document-vector-pipeline/tree/main/AzureSQL/csharp`.

Provide Building Blocks

We set out from the beginning to **not** be an AI model hosting system. We wanted our customers to have the widest choices of which AI models they wanted to use whether they be all on-premises or connected to a cloud. So we built a series of T-SQL and engine capabilities that are *building blocks* for AI applications that can use vector search *with their data* inside SQL Server. We wanted to make sure we could support modern AI frameworks like LangChain, Semantic Kernel, Entity Framework (EF) Core, and Model Context Protocol (MCP) Servers.

Promote Security and Scalability

Almost every customer I talked to about AI with their data says security is their biggest concern. We wanted to build capabilities inside our database engine because we knew our customers trusted the *security blanket* of SQL Server. SQL Server is already designed to be scalable, so we knew we could rely on the built-in scale of the database engine to support the needs of any AI application wanting to use vector search.

Overcome Complexity

Our team has a rich history of leveraging the power of the familiar T-SQL language to support a wide variety of features. So why not use T-SQL for AI? T-SQL has so many advantages including allowing us to provide security for all the capabilities of AI inside the engine.

What Is AI-Ready?

What does AI-ready mean for SQL Server 2025? It means providing the ability to access AI models ground or cloud and use a new **vector data type** to execute vector searching on your data inside the security of the SQL Server engine. Let's look at the new features in SQL Server 2025 that make this possible.

As you read this section, for a complete list of features for SQL Server 2025 for AI, take a look at `https://learn.microsoft.com/sql/sql-server/what-s-new-in-sql-server-2025#ai`.

Note I will mention the need for the database scoped configuration option PREVIEW_FEATURES in this chapter for features like vector indexes and VECTOR_SEARCH. However, at GA for SQL Server 2025, you will have full production support for the vector data type, model definitions, and vector searching with the VECTOR_DISTANCE function. All built-in to SQL Server 2025.

Vector Data Type

As I described in Chapter 3, a vector is an ordered array of floating-point numbers. SQL Server has a new data type called **vector(n)** where n represents the number of dimensions the vector can store. Because SQL Server can natively support the storage, management, and query of vectors, we can now officially call it a *vector store*. While there are solutions available today that are just a vector store, many popular databases are including vector store support inside the product like SQL Server 2025 (and Azure SQL and SQL Database in Fabric).

Vectors are stored in an optimized binary format but are exposed as JSON arrays for convenience when you try to extract them from a SELECT statement. Each element of the vector is stored as a single-precision (4-byte) floating-point value.

Dimensions

Dimensions are the number of floating-point numbers a given word, phrase, sentence, or set of text represents. Think of these numbers all in a multidimensional space. The closer the distance of two sets of dimensions is, the more similar they are in meaning. In general, you can think of the higher the dimensions, the better the results you might get, but in many scenarios a large number of dimensions are not required.

SQL Server 2025 today supports a maximum of 1,998 dimensions for a vector type (we are looking to expand this in the future). You need to specify the dimensions for your vector type that align with the dimensions of the embedding model you choose. And the issue is that you won't realize this until you try to use a function like AI_GENERATE_EMBEDDINGS, which could yield an error like this:

```
Msg 42204, Level 16, State 2, Line 5
The vector dimensions 500 and 1024 do not match.
```

Choosing a model with the right dimensions within what we support is a good exercise for you to research. However, we have seen scenarios where you don't necessarily need very large dimensions. Read more in an article my colleague Davide Mauri wrote at `https://devblogs.microsoft.com/azure-sql/embedding-models-and-dimensions-optimizing-the-performance-resource-usage-ratio`.

Note You will see below that the recommended use for a vector type is to store embeddings from an embedding AI model using the T-SQL function **AI_GENERATE_EMBEDDINGS**. However, a vector type can take any array of floating-point numbers that matches your dimensions. So technically a data scientist could store feature vectors with it. And any application can generate embeddings using an outside method and store them in the vector data type.

Driver Support

The vector data type is a binary format *inside* SQL Server but presented to applications as JSON text. We have updated both the TDS protocol and drivers to transmit vector data more efficiently in binary format and present them to applications as native vector types. This approach reduces payload size, eliminates the overhead of JSON parsing, and preserves full floating-point precision. As a result, it improves both performance and accuracy when working with high-dimensional vectors in AI and machine learning scenarios (this is right from our docs at `https://learn.microsoft.com/sql/t-sql/data-types/vector-data-type#compatibility`).

We have also updated specific drivers to support the TDS protocol changes including **Microsoft.Data.SqlClient** and the **Microsoft JDBC Driver for SQL Server** using new types. Other and older drivers "just work" but see a vector as nvarchar(max).

Model Definitions

You must have an AI model available to generate embeddings to store in the new vector data type. And we knew we could already support the use of **sp_invoke_external_rest_endpoint** to send text to an embedding model and convert the JSON result into our vector type. However, this meant that you had to call this system procedure for every row in your table to generate a vector. My early examples used a server cursor (I know, please close your eyes when you read this) to iterate through a table and call this procedure one row at a time. One important aspect to this scenario is that it requires a REST endpoint over **a secure HTTPS protocol**.

CREATE EXTERNAL MODEL Is Born

Our team spent a long time thinking through a solution for this. We studied the competition and what others were doing in the industry. One principle we strove from the very beginning was this: *we will not load AI models into the database engine*. Even though as I described in Chapter 3 that AI models are not executable programs, we all just felt that this was a deal breaker for many enterprise customers. Therefore, after a long debate on the topic, we landed on creating a new T-SQL syntax to allow users to define a model definition so that metadata about the model and how to access it would be stored inside a system table.

Thus was born the **CREATE EXTERNAL MODEL** T-SQL statement, which is a database scoped statement. We tasked Brian Spendolini, Senior Product Manager, and the product owner for sp_invoke_external_rest_endpoint to come up with a solution. As Brian tells it, *"Around May of 2024, I remember being asked to come up with a way that we could connect to AI endpoints from SQL Server. My first design was awful; it was just a giant object that had six fields that took in random JSON strings. As the design evolved, we knew we needed to compromise between ease of use and customization so that it could evolve with the light speed pace that AI was and is still growing. It also had to be secure; the model definition could not be changed or altered to point to endpoints that contained biased or nefarious models but could be simply provided to developers to use. I remember the final design review meeting very well. I was on 'vacation' in Lake Placid. I had COVID, isolated in a basement, and the Wi-Fi was spotty. Regardless, the final design was presented and shortly after passed to the engineers to start coding."*

The **CREATE EXTERNAL MODEL** T-SQL statement has this structure from the documentation at `https://learn.microsoft.com/sql/t-sql/statements/create-external-model-transact-sql`:

```
CREATE EXTERNAL MODEL external_model_object_name
[ AUTHORIZATION owner_name ]
WITH
  (    LOCATION = '<prefix>://<path>[:<port>]'
     , API_FORMAT = '<OpenAI, Azure OpenAI, etc>'
     , MODEL_TYPE = EMBEDDINGS
```

```
    , MODEL = 'text-embedding-model-name'
    [ , CREDENTIAL = <credential_name> ]
    [ , PARAMETERS = '{"valid":"JSON"}' ]
    [ , LOCAL_RUNTIME_PATH = 'path to the ONNX runtime files' ]
);
```

Note SQL Server 2025 also supports an **ALTER EXTERNAL MODEL** T-SQL statement to modify an existing definition which you can read about at `https://learn.microsoft.com/sql/t-sql/statements/alter-external-model-transact-sql`. Be careful with this statement. For example, you could change the model definition but you should consider regenerating your embeddings to match this new definition if you make this change.

Let's unpack the statement further:

external_model_object_name

This is an object identifier (like a table) that can be used with other T-SQL statements like **AI_GENERATE_EMBEDDINGS.**

AUTHORIZATION owner_name

This allows you to assign ownership to a user or role that owns the model definition and can then decide who can use the model. This is not required, and if not specified the user who executes the T-SQL statement is assumed to be the owner. A user must have CREATE EXTERNAL MODEL database permission to run this statement. A sysadmin is the only security principal with default permission for this statement. Any other user, including any db_owner user, must be granted specific permission for this. This adds to the security posture of using AI with SQL Server 2025.

LOCATION = '<prefix>://<path>[:<port>]

The location is the URL of the REST endpoint for the AI model hosting service whether it be on-premises or in a cloud provider. The format of this URL depends on the API_FORMAT. It is also possible that this location can be a file path for an ONNX model. You can learn more about this in the section titled "Local ONNX Support" later in this chapter. You will see examples of what the URL endpoint can be later in this chapter.

One of the early decisions we landed on is that we require the location **to support HTTPS** with no exceptions to ensure data is always encrypted. Some model hosting services don't natively support HTTPS, so you may need to use a proxy service that supports HTTPS and redirect this to your model hosting service.

API_FORMAT = '<OpenAI, Azure OpenAI, etc>'

There was a ton of research behind API_FORMAT. We wanted to take out complexity of how you had to define this endpoint compared with using **sp_invoke_external_rest_endpoint**, which had to be very specific to the URL syntax. By allowing you to tell us which API_FORMAT, we could take away some of the complexity.

We spent countless hours debating what REST protocols we should support. **Azure OpenAI** was a natural choice (`https://learn.microsoft.com/azure/ai-foundry/openai/reference`) for Azure AI Foundry models based on Azure OpenAI.

We also knew we should support **Ollama** as it was a very popular on-premises open source service (`https://deepwiki.com/ollama/ollama/3-api-reference`).

We also knew that many customers would have a subscription and use OpenAI (`https://github.com/openai/openai-openapi`). What we discovered is that many other hosting services, on-premises and in the cloud, were using the OpenAI standard. Therefore, we added the **API_FORMAT = OpenAI**.

Note The reason why Azure OpenAI is different from OpenAI is that the Azure OpenAI has specific requirements, security, and lower latencies that are unique to using OpenAI in Azure AI Foundry. Here is where it gets confusing. Azure AI Foundry supports other models than Azure OpenAI, and for those you will use the **API_FORMAT = 'OpenAI'**.

In addition, as you will learn in the section titled "Local ONNX Support," we also support a protocol to use a local ONNX model **that does not use REST** but it's not loaded in the SQL Server engine.

MODEL_TYPE = EMBEDDINGS

Early designs included a CREATE EXTERNAL EMBEDDING MODEL statement, but we decided to use a MODEL_TYPE. The only model supported today is embeddings, but we have reserved a design that allows us to support other model types in the future.

MODEL = 'text-embedding-model-name'

This is the name of the embedding model you are using in your hosted service, deployment, or ONNX runtime.

[, CREDENTIAL = <credential_name>]

This is an optional parameter that supports the name of a DATABASE SCOPED CREDENTIAL if your hosting service requires an authentication. This could be an API_KEY or an Azure Managed Identity. Many local hosting services can support through a proxy a self-signed local certificate, which does not require authentication from SQL Server (because any program on the computer or VM is automatically authenticated).

[, PARAMETERS = '{"valid":"JSON"}']

We wanted to provide a convenient method for you to use different parameters for an AI model to change its behavior, and many model types accept various parameter values. For embedding models one of these is **dimensions**. For example, you could be using an embedding model that is designed for 2,048 dimensions, but that exceeds the current maximum for the SQL Server 2025 vector type. Some models allow you to tune the maximum dimensions to use so you could use a parameter like this:

```
'{ "dimensions": 1536 }'
```

This is important because you might find the exact embedding model you like but the dimensions are larger than our current vector type support. Consult the model documentation to see if this parameter is supported.

[, LOCAL_RUNTIME_PATH = 'path to the ONNX runtime files']

This value is optional but required if you are using an ONNX model. I'll discuss this further later in the chapter under the section titled "Local ONNX Support."

To use CREATE EXTERNAL MODEL, you must enable the server configuration value **external rest endpoint enabled.**

System View

All model definitions in your database can be discovered through the system view **sys.external_models** (`https://learn.microsoft.com/en-us/sql/relational-databases/system-catalog-views/sys-external-models-transact-sql`).

Embedding Generation

You have your model definition in place for embeddings. Now what do you do? You can generate embedding referencing the external model identifier using the T-SQL function **AI_GENERATE_EMBEDDINGS** (`https://learn.microsoft.com/sql/t-sql/functions/ai-generate-embeddings-transact-sql`).

This function takes as parameters a **source**, which is any text expression or column that is **nvarchar**, **varchar**, **nchar**, or **char.** This could be an expression that is a combination of text columns.

You also specify the external model identifier with **USE model_identifier**. This must be a fixed name and cannot be a variable. In addition, you can choose an optional JSON parameter with each call.

Since this is a system T-SQL function, you can use this statement in the context of a SELECT, INSERT..SELECT, or UPDATE (so no more cursors). However, there is no batching concept, so internally we process one row at a time and the execution is synchronous.

The result of this function is a JSON array of embeddings. But if you use a T-SQL INSERT..SELECT or UPDATE into a vector data type, we automatically convert the data into the vector data type binary format.

If you encounter errors, use the XEvent **ai_generate_embeddings_summary** to help diagnose any problems.

Text Chunking

I mentioned the concept of token limits for AI models in Chapter 3. Embedding models have token limits on the text you can send to generate embeddings. In some cases, an AI hosting service using an embedding model will fail if you exceed this limit. For example, with Azure OpenAI you might see this error:

```
'This model's maximum context length is 8191 tokens, but you requested 8193
tokens (8193 in your prompt; 0 for the completion)'
```

Tokens for embedding models are approximately **1 token ≈ 4 characters in English** (average). The popular model from OpenAI, text-embedding-ada-002, supports 8,192 tokens or ~12–16 thousand characters. Models have different token limits, and unfortunately some hosting services like Ollama will not return an error but silently truncate the text sent to the model (yikes).

In either case if you have text data in a column of your table that exceeds this limit, you will need to possibly use a technique called "chunking" or break up your text to use with an embedding model.

SQL Server 2025 comes with a T-SQL function to help called **AI_GENERATE_CHUNKS.** This function will help you break up your text into chunks to send to an embedding model. At the time of the writing of this book, this function supports a "fixed" set of text to chunk your overall text data and requires the PREVIEW_FEATURES database scoped configuration option. It is very possible this option will not be required at GA. However, we may explore other options that will require this option in the future with cumulative updates that can use more natural techniques like "paragraphs" using full-text search. Learn more how to use chunking at `https://learn.microsoft.com/sql/t-sql/functions/ai-generate-chunks-transact-sql`.

Searching with VECTOR_DISTANCE

At this point you are ready to go. You can use a T-SQL function called **VECTOR_DISTANCE** to find the most similar rows to your prompt using the vector type. The VECTOR_DISTANCE function, `https://learn.microsoft.com/sql/t-sql/functions/vector-distance-transact-sql`, uses the concept of k-nearest neighbor or **KNN**.

The concept of VECTOR_DISTANCE is to compare two vector values and provide the distance between them using the options of

cosine - Cosine distance

euclidean - Euclidean distance

dot - (Negative) Dot product

The concept is that the closer the distance, the more similar the vector numbers are. And since you are using embeddings for vectors, which have semantic meaning, the closer the distance, the more *similar* the text.

So which "distance" method should you choose? Here is where using AI starts to get a bit more complex, but I'll make it simple for you. Most modern AI models return embeddings as *normalized.* If that is the case, then the **cosine** option should work just fine. The other options may be better if your embeddings are not normalized.

Great. How do I know if my embeddings are normalized? The embedding model and hosting service documentation should say this, but we also have T-SQL functions to help:

VECTOR_NORM - Use VECTOR_NORM to take a vector as an input and return the norm of the vector (which is a measure of its length or magnitude).

VECTOR_NORMALIZE - Use VECTOR_NORMALIZE to take a vector as an input and return the normalized vector.

Tip Use a model where vectors are normalized and use cosine. It has worked very well for me, but there are deep AI experts that may want to use the other distance options.

Returning the distance is great, but how do use VECTOR_DISTANCE to return a "search"? A simple example is in our documentation like this:

```
DECLARE @v AS VECTOR (1536);
SELECT @v = title_vector
FROM [dbo].[wikipedia_articles]
WHERE title = 'Alan Turing';

SELECT id,
       title,
       VECTOR_DISTANCE('cosine', @v, title_vector) AS distance
FROM [dbo].[wikipedia_articles]
WHERE VECTOR_DISTANCE('cosine', @v, title_vector) < 0.3
ORDER BY distance;
```

Remember that using this function is a scan and could even result in a spool operator, which will require space in tempdb. It works great for smaller sets of data, but what if you have millions of vectors to search?

Vector Index

VECTOR_DISTANCE is a precise lookup requiring a scan of your table. And we are SQL Server, so why not use an index?

Creating a b-tree index on SQL Server data is our jam. We know how to make searching very fast on all of your data in SQL Server. But how do you create an index on vectors? And how do you make it efficient for very large sets of vectors especially if they don't all fit in memory? This is where a **DiskANN index** comes in.

DiskANN Indexes

Most common vector indexes are based on the principle of **Approximate Nearest Neighbor (ANN).** ANN-based indexes are designed to quickly find vectors that are close to a query vector, but it does not guarantee exact results. Instead, it trades accuracy for speed and scalability, which is essential when dealing with millions or billions of high-dimensional vectors.

One of the most popular ANN-based vector indexes is **Hierarchical Navigable Small World (HNSW).** HNSW indexes are based on a graph structure. Most HNSW indexes are “in-memory” where each node in the graph as well as each edge between nodes are stored in memory.

In 2019, Microsoft research along with academia experts proposed a new method for vector indexes called **DiskANN** (`https://suhasjs.github.io/files/diskann_neurips19.pdf`). The concept was to use fast SSDs to not require a vector index be fully “in-memory.” This became an open source project in 2021 (`https://github.com/microsoft/DiskANN`). Now it is almost a “Microsoft standard” as many Microsoft products have adopted this vector index type.

SQL Server DiskANN Index and VECTOR_SEARCH

SQL Server 2025 now supports a DiskANN index via the **CREATE VECTOR INDEX** statement (`https://learn.microsoft.com/sql/t-sql/statements/create-vector-index-transact-sql`).

With a vector index in place, now you can use the **VECTOR_SEARCH** T-SQL function (`https://learn.microsoft.com/sql/t-sql/functions/vector-search-transact-sql`) to search through the index based on an input vector (say an embedding generated from a prompt).

While this is an approximate search, my experience is this is a very excellent approach for vector searching especially for very large sets of vectors.

You will see an example of using a vector index and search in the examples later in this chapter.

Limitations

At the time of the writing of this book, the database scoped configuration option PREVIEW_FEATURES is required to use CREATE VECTOR INDEX and VECTOR_SEARCH. After the GA of SQL Server 2025, we will iterate and improve vector indexing to make it "GA"-ready through cumulative updates.

Our intention is that we want you to be able to develop and test using vector indexes as we improve them. This is because at the time of the writing of this book, vector indexes have these limitations:

- **A table with a vector index becomes read-only.**

  This is the biggest limitation I know customers will be frustrated by. This is similar to when columnstore indexes first were available but we quickly were able to make the table updatable.

- **The table must have a single-column, integer, primary key clustered index.**
- **Vector indexes aren't replicated to subscribers.**
- **VECTOR_SEARCH is post-filter only.**

  Vector search happens before applying any predicate. Additional predicates are applied only after the most similar vectors are returned.

- **VECTOR_SEARCH can't be used in views.**

  The goal is to remove all of these limitations when vector indexes and VECTOR_SEARCH don't require the PREVIEW_FEATURES option.

There are also performance improvements we are laser-focused on to improve the speed of building vector indexes. One of our goals is to ensure we perform well with the popular open source benchmark `https://github.com/MageChiu/VectorDBBench`.

Learn More

There is a great FAQ on vectors that might answer questions you still may have at `https://learn.microsoft.com/en-us/sql/relational-databases/vectors/vectors-faq`.

Vector Architecture

I feel nothing is better to describe how something works than a great visualization. Figure 4-1 represents the SQL vector architecture and flow of using it with SQL Server 2025.

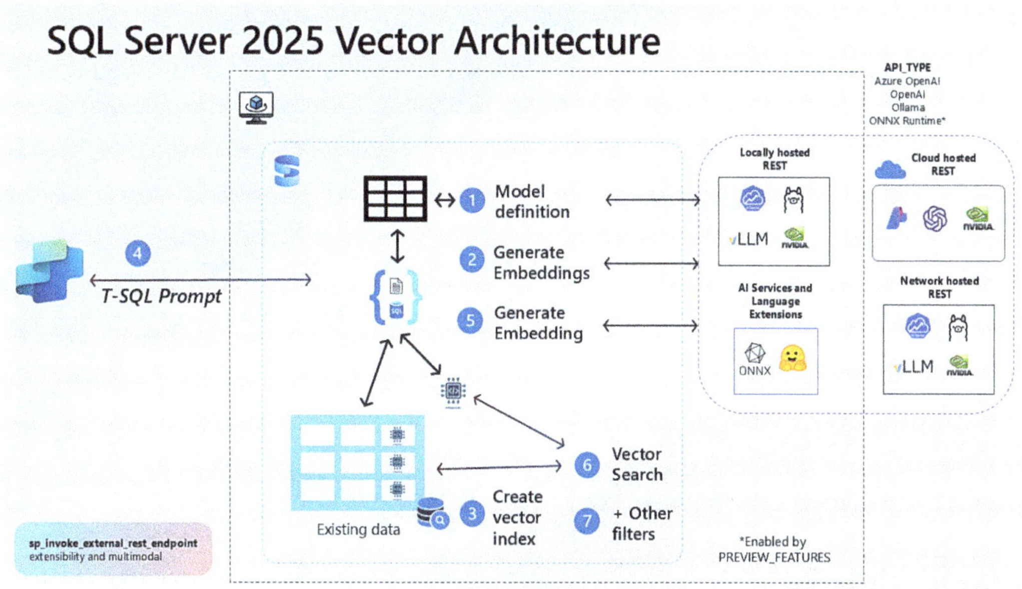

Figure 4-1. The SQL Server vector architecture

Let's look at each step. Keep these steps in mind as you will use them in an example later in this chapter.

1. Model Definition

Use CREATE EXTERNAL MODEL for the embedding model of your choice. This can be models hosted locally via REST through services such as KServe, Ollama, vLLM, or NIM. They can also be accessed locally through ONNX.

You can also use REST to access AI models on your local network though the same set of services. Or you can use cloud providers such as Azure AI Foundry, OpenAI, or NVIDIA.

This is one of the best stories of SQL Server and AI. You use the model of your choice, locally, on a local network, or with a cloud provider, secure and isolated from the SQL Server engine.

We support API_PROMPT types for **Azure OpenAI, OpenAI, Ollama,** and **ONNX Runtime.** These match the REST protocols for the most popular embedding models.

Note At the time of the writing of this book, I have been working closely with the Foundry Local team to support embedding models. Once that happens it will provide a nice option for local AI models for Windows Server customers. See more at `https://learn.microsoft.com/azure/ai-foundry/foundry-local`.

---

2. Generate Embeddings

Use the model definition to generate embeddings from your existing text data using **AI_GENERATE_EMBEDDINGS** and store these in a **vector** data type column in your existing tables or a new table that can join with your text data.

3. Create a Vector Index

Optionally create a vector index on your vector data type column.

4. T-SQL Prompt

Use T-SQL to submit a prompt to your database through, for example, a stored procedure as nvarchar.

5. Generate Prompt Embedding

In the procedure use **AI_GENERATE_EMBEDDINGS** to generate an embedding for the prompt using the same model definition.

6. Vector Search

Use the generated embedding from the prompt to perform a vector search using VECTOR_DISTANCE or VECTOR_SEARCH.

7. Other Filters

Combine your vector search with other WHERE clauses to filter your search. One interesting twist is the concept of **re-ranking.** Vector search is amazing, but in some scenarios it might be desirable to apply a more sophisticated or *domain-specific scoring function to reorder the top candidates.*

In comes SQL Server full-text search. You can use SQL Server full-text search (FTS) to (re)rank the top-K candidates you got from a vector search—or run both FTS and vector searches in parallel and fuse their rankings with a simple, robust method like **Reciprocal Rank Fusion (RRF).** SQL Server's FTS already gives you a high-quality *BM25-style relevance* score via CONTAINSTABLE/FREETEXTTABLE, while the new vector features (Azure SQL + SQL Server 2025 preview) give you exact and approximate semantic similarity in-database. Combine them, and you get the best of lexical + semantic. And of course Davide Mauri has an excellent example of how to use full-text search with vector search for re-ranking at `https://devblogs.microsoft.com/azure-sql/enhancing-search-capabilities-in-sql-server-and-azure-sql-with-hybrid-search-and-rrf-re-ranking`.

Extending with sp_involve_external_rest_endpoint

If for some reason our CREATE EXTERNAL MODEL types don't meet your needs, you can always use sp_invoke_external_rest_endpoint to communicate with any embedding model that can be accessed via REST.

This architecture represents the hard work and incredible efforts by so many people. I asked Alexey Eksarevskiy, Principal Software Engineer and someone so instrumental for this release, for his thoughts: *"Since we already had support for sp_invoke_external_rest_endpoint() in the SQL Engine code, it was natural for us to leverage the same code to implement REST calls into AI model endpoints. However, we looked further and saw an opportunity to benefit from other capabilities that SQL has to offer, for example, ability to create metadata-only objects and leverage comprehensive security model (including separation of roles between a "model admin" and developer), ability to create schema-bound views, etc. Abstracting out all the details necessary to connect to a model in an external model object makes it simple to write ai_generate_embeddings() calls and update any model details without touching that code (even switch the model completely). Another nice feature of external model as an abstraction allows us to add support for more model endpoints in the future, both hosted remotely or locally, add more model types, or even provide an opportunity for the developer to define the 'API format' specifics in the model itself, in order to be able to keep up with the rapidly changing landscape in the AI space. We were happy to see more model providers standardize on the OpenAI REST API (that's why we switched from the concept of "model provider" we originally had to API_FORMAT), but at the same time it is hard to predict what other formats may emerge in the future."*

Why Enterprise?

I remember vividly talking about our theme of AI-ready enterprise for SQL Server 2025 as we prepared to launch the product in November 2024. I pushed hard for the term *enterprise* in the theme. My argument is that customers are concerned about the security of using AI but trust the enterprise capabilities of SQL Server. Consider these important security principles using SQL Server 2025 and AI.

You Control All Access with SQL Security

I've had someone ask me when I presented SQL Server 2025 and AI "how do I turn if off." The better question maybe "how do I turn it on." Consider this:

- You cannot use AI models inside SQL Server unless you first turn on the option **external rest endpoint enabled** for REST or **external AI runtimes enabled** for ONNX (which also requires you to choose a specific feature).
- CREATE EXTERNAL MODEL requires specific permissions to execute.
- All AI capabilities are through T-SQL, and they are all integrated with the enterprise-grade SQL security model.

You Control Which AI Models to Use

You decide which AI model to use ground to cloud. Stay completely on-premises locally or on your network. Or go hybrid and use cloud-hosted providers. You are in complete control of which AI model to use with all the popular models supporting Azure OpenAI, OpenAI (and OpenAI compatible), Ollama, and ONNX runtime.

AI Models Ground and/or Cloud Isolated from SQL

As you choose your model ground or cloud, remember that we **do not load AI models** in the SQL Server engine. All models run isolated from the SQL Server engine.

Use RLS, TDE, and DDM

Use popular SQL Server security features like Row-Level Security (RLS), Transparent Data Encryption (TDE), and Dynamic Data Masking to secure your data.

Track Everything with SQL Server Auditing

Since all of the AI features use T-SQL, SQL Server Auditing can track all access to AI models within the engine, all access to vector data types, and all vector searches.

Ledger for Chat History and Feedback

Some AI applications that use a chat prompt approach like to offer customers the ability to store history of chat conversations and track feedback on quality. If you desire to do this, consider using SQL Server Ledger (`https://aka.ms/sqlledger`) for an immutable and tamper-evident approach for this important information.

Getting Started with Vectors

Let's use the vector architecture as a guide to try an example (even on your laptop). Here is the problem we are trying to solve. The AdventureWorks database has product descriptions in the **Production.ProductDescription** table. We would like to provide a rich search experience for product descriptions to support a RAG pattern we are building. We will use the following prompt example: **"Show me stuff for extreme outdoor sports."** You will find out that the words "stuff" and "extreme" are not in any product description, but AI models can help us find similar results just like we understand these words as humans.

Prerequisites

If you read through Chapter 3 and followed the examples, you already have Ollama and caddy (for a proxy) installed. If not, go back to Chapter 3 and follow the instructions in the section title "Let's REST with AI." Make sure Ollama and caddy are both running.

All the scripts for this example can be found in the **ch4_AI\ollama_ai** folder.

1. In Chapter 3 we used the caddy program for the self-signed certificates that were valid since curl was run under your user context. Since SQL Server runs as a service, you will need to import the certificate from caddy into the Local Computer certificate store if you are using Windows Server.
   Locate the caddy's certificate from %APPDATA%\Caddy\pki\authorities\local\root.crt.
   - Run **certlm.msc**.
   - Navigate to **Trusted Root Certification Authorities ➤ Certificates**.
   - Right-click ➤ **All Tasks ➤ Import**.
   - Select root.crt and complete the wizard.
   - **Restart the SQL Server service** to ensure it picks up the updated trust store.

Note Here is an example for importing the certificate on Linux: `https://documentation.ubuntu.com/server/how-to/security/install-a-root-ca-certificate-in-the-trust-store`. Here is a possible resource for you to do this with containers: `https://www.baeldung.com/ops/docker-container-import-ssl-certificate`.

---

2. Install SQL Server 2025 Developer (Enterprise) Edition. You will need to choose the full-text search feature during installation.
3. For the purposes of this example, log in to SQL Server as a sysadmin to avoid having to set any special permissions.

4. Download the same database **AdventureWorks** from `https://github.com/Microsoft/sql-server-samples/releases/download/adventureworks/AdventureWorks2022.bak`.

5. Restore the database using the script **restore_adventureworks.sql** (you may need to edit the file paths for the backup and/or database and log files).

6. Enable REST API for model definitions using the script **enablerestapi.sql**. When models are accessed, SQL Server implicitly uses REST, so this enables the **external rest endpoint enabled** configuration option.

7. Create a full-text index using the script **createft.sql**. I'll use this to compare full-text index searching with a vector search.

8. Enable the database option **PREVIEW_FEATURES** for vector index using the script **enable_preview_features.sql**.

Note Because caddy as a proxy uses a self-signed certificate on the local VM or machine, you will not need a database scoped credential like you would for a hosted service on a network or cloud service.

Try Existing Search Methods

Before you see the vector search experience, let's try a "prompt" we will use for a vector search using existing SQL Server "search" methods.

1. Load the script **search_productdescription.sql** that runs the following T-SQL statements:

```
USE AdventureWorks;
GO
SELECT * FROM Production.ProductDescription
WHERE Description LIKE '%Show me stuff for extreme outdoor
sports%'
GO
```

```
SELECT * FROM Production.ProductDescription
WHERE CONTAINS(Description, '"Show me stuff for extreme outdoor
sports"');
GO
SELECT * FROM Production.ProductDescription
WHERE FREETEXT(Description, 'Show me stuff for extreme outdoor
sports');
GO
```

The first two queries return 0 rows. The third query does return two rows, but I suspect we can do better (plus our database may contain multiple languages).

Step 1: Create a Model Definition

Per the visual in Figure 4-1, the first step is to create a model definition. If you look back at Chapter 3 in the section titled "Use REST with Ollama to Do a Vector Search," the URL for an embedding model with Ollama is **https://localhost/api/embeddings**. This is the endpoint for Ollama for a single embedding. For SQL Server, we need to use the "batch" endpoint, which is **/api/embed**.

1. Create an external model definition by executing the script **create_external_model.sql**. This script contains the following T-SQL statements:

```
USE [AdventureWorks];
GO
IF EXISTS (SELECT * FROM sys.external_models WHERE name =
'MyOllamaEmbeddingModel')
DROP EXTERNAL MODEL MyOllamaEmbeddingModel;
GO
-- Create the EXTERNAL MODEL
CREATE EXTERNAL MODEL MyOllamaEmbeddingModel
WITH (
      LOCATION = 'https://localhost/api/embed',
      API_FORMAT = 'Ollama',
      MODEL_TYPE = embeddings,
```

```
    MODEL = 'mxbai-embed-large');
GO
SELECT * FROM sys.external_models;
GO
```

The final query should show a single row for the new model definition. Remember this is just metadata. We haven't loaded an AI model into SQL Server. In this case we are using the embedding model we pulled in Chapter 3 called **mxabi-embed-large**.

Tip As I was writing this chapter, my colleague Muazma Zahid told me about Ollama updates that include a new small embedding model by Google. You can see it at `https://ai.google.dev/gemma/docs/embeddinggemma`. So the tip is to stay up to date with the latest in embedding model technology.

Step 2: Create a Table to Store Vectors and Generate Embeddings

With the model definition in place, we can use the model to generate embeddings and store them into a vector type.

1. Load and execute the script **embeddingtable.sql.** This script contains the following T-SQL statements:

```
USE AdventureWorks;
GO
-- Create a new table to store embeddings
--
DROP TABLE IF EXISTS Production.ProductDescriptionEmbeddings;
GO
CREATE TABLE Production.ProductDescriptionEmbeddings
(
  ProductDescEmbeddingID INT IDENTITY NOT NULL PRIMARY KEY
  CLUSTERED,
  ProductID INT NOT NULL,
```

```
  ProductDescriptionID INT NOT NULL,
  ProductModelID INT NOT NULL,
  CultureID nchar(6) NOT NULL,
  Embedding vector(1024)
);
```

This table is what I call a "join" table as it contains keys to join to other tables in the database including product descriptions. A clustered unique index is required (for now—this could change in the future) to create a vector index. Notice this important column:

Embedding vector(1024)

This is a column for the new vector type. The number is the dimensions of the vector. I explained the concept of dimensions earlier in the chapter describing the new vector type. The dimension number specified here cannot exceed the maximum number of dimensions supported by the model. You can read at `https://www.mixedbread.com/docs/models/embedding` that this model supports a maximum of 1,024 dimensions.

You don't have to create a separate table for a vector type, but I like to keep it separate from the original table that includes my text data.

2. Load and execute the script **generate_embeddings.sql** that contains the following T-SQL statements:

```
USE AdventureWorks;
GO
-- Populate rows with embeddings
-- Need to make sure and only get Products that have ProductModels
INSERT INTO Production.ProductDescriptionEmbeddings
SELECT p.ProductID, pmpdc.ProductDescriptionID, pmpdc.
ProductModelID, pmpdc.CultureID,

AI_GENERATE_EMBEDDINGS(pd.Description USE MODEL
MyOllamaEmbeddingModel)
```

```
FROM Production.ProductModelProductDescriptionCulture pmpdc
JOIN Production.Product p
ON pmpdc.ProductModelID = p.ProductModelID
JOIN Production.ProductDescription pd
ON pd.ProductDescriptionID = pmpdc.ProductDescriptionID
ORDER BY p.ProductID;
GO
```

This is where the magic happens:

AI_GENERATE_EMBEDDINGS(pd.Description USE MODEL MyOllamaEmbeddingModel)

Since **AI_GENERATE_EMBEDDINGS** is a T-SQL function, you can use this naturally to process rows for INSERT...SELECT or UPDATE. This is a "one row at a time" feature for now. There is no batching concept today. So for each row, we will use the model definition to send a REST call to the endpoint (in my case to caddy, which is redirected to Ollama). The model will respond with embedding values. We automatically take this ordered array of floating-point numbers (as JSON) and put them in a binary format inside the vector type.

In this scenario, I was able to generate 1,764 rows of embeddings in about one minute. I didn't have to use any text chunking technique here so that could change your expectations to "load" your embeddings. However, I will admit my laptop has a GPU. I tried this on a laptop without a CPU, and it took significantly longer.

Important In this example I don't use text chunking with AI_GENERATE_CHURNKS because my data in the product description does not exceed token limits for my AI embedding models.

---

3. Explore what vectors look like by loading and executing the script **explore_embeddings.sql** that has the following T-SQL statement:

```
USE AdventureWorks;
GO
SELECT TOP 20 p.ProductID, p.Name, pd.Description, pde.Embedding
FROM Production.ProductDescriptionEmbeddings pde
JOIN Production.Product p
ON pde.ProductID = p.ProductID
JOIN Production.ProductDescription pd
ON pd.ProductDescriptionID = pde.ProductDescriptionID
GO
```

This query will show the embeddings for a sample of product descriptions. You can see the values look like an array of floating-point numbers. We always show the values this way when you query the column, but internally they are stored in a binary format. My results look like Figure 4-2.

	ProductID	Name	Description	Embedding
1	680	HL Road Frame - Black, 58	لقد تم صناعة هيكل دراجتنا الألومنيوم الأخف وزنًا والأعلى ج...	[-2.3432977e-002,-3.7991278e-002,-1.8213709e-003,...
2	706	HL Road Frame - Red, 58	מסגרת האלומיניום הקלה והאיכותית ביותר שלנו עשויה ...	[-1.5711028e-002,-1.1890588e-002,-2.7214695e-002,...
3	706	HL Road Frame - Red, 58	เฟรมอลูมิเนียมคุณภาพสูงสุดและน้ำหนักเบาที่สุด สร้างจากอัลล...	[-5.9569497e-003,2.4337566e-003,3.7983764e-002,9....
4	706	HL Road Frame - Red, 58	Notre cadre en aluminium plus léger et de qualité sup...	[9.8592052e-003,1.9109076e-002,-2.5642809e-002,2....
5	706	HL Road Frame - Red, 58	لقد تم صناعة هيكل دراجتنا الألومنيوم الأخف وزنًا والأعلى ج...	[-2.3432977e-002,-3.7991278e-002,-1.8213709e-003,...
6	706	HL Road Frame - Red, 58	Our lightest and best quality aluminum frame made fr...	[1.7936709e-003,-1.5676210e-002,-2.8966570e-002,...
7	707	Sport-100 Helmet, Red	Universal fit, well-vented, lightweight , snap-on visor.	[-3.6746383e-002,-3.1093501e-003,-1.5409609e-002,...
8	707	Sport-100 Helmet, Red	ملائمة بشكل عام، وجيدة التهوية، وخفيفة الوزن بقناع واق م...	[-1.9980656e-002,-3.5985049e-003,-8.3055878e-003,...
9	707	Sport-100 Helmet, Red	Légère, aérée, taille unique, avec une visière amovible.	[-2.7858917e-002,-2.3723708e-002,-4.5293834e-002,...
10	707	Sport-100 Helmet, Red	แว่นกันลมขนาดสากล ระบายอากาศได้ดี น้ำหนักเบา	[1.1851110e-002,1.9434173e-002,3.9684977e-002,6....
11	707	Sport-100 Helmet, Red	מידה אוניברסלית, מאווררת היטב, קלת-משקל, עם מצחיי...	[-2.8125532e-002,-4.6258837e-002,-3.4795206e-002,...
12	707	Sport-100 Helmet, Red	通用型透气良好且轻便, 带有自合型帽沿.	[-1.9060943e-002,3.4912003e-003,-1.3274158e-002,...
13	708	Sport-100 Helmet, Black	通用型透气良好且轻便, 带有自合型帽沿.	[-1.9060943e-002,3.4912003e-003,-1.3274158e-002,...
14	708	Sport-100 Helmet, Black	מידה אוניברסלית, מאווררת היטב, קלת-משקל, עם מצחיי...	[-2.8125532e-002,-4.6258837e-002,-3.4795206e-002,...
15	708	Sport-100 Helmet, Black	แว่นกันลมขนาดสากล ระบายอากาศได้ดี น้ำหนักเบา	[1.1851110e-002,1.9434173e-002,3.9684977e-002,6....
16	708	Sport-100 Helmet, Black	Légère, aérée, taille unique, avec une visière amovible.	[-2.7858917e-002,-2.3723708e-002,-4.5293834e-002,...
17	708	Sport-100 Helmet, Black	ملائمة بشكل عام، وجيدة التهوية، وخفيفة الوزن بقناع واق م...	[-1.9980656e-002,-3.5985049e-003,-8.3055878e-003,...
18	708	Sport-100 Helmet, Black	Universal fit, well-vented, lightweight , snap-on visor.	[-3.6746383e-002,-3.1093501e-003,-1.5409609e-002,...
19	709	Mountain Bike Socks, M	Combination of natural and synthetic fibers stays dry a...	[-2.6916960e-002,6.6534318e-002,1.0017234e-002,2....
20	709	Mountain Bike Socks, M	تركيبة من الفيبر الطبيعي والصناعي تظل محتفظة بجفافها ...	[-3.4392778e-002,-2.7904170e-002,-1.3461495e-002,...

***Figure 4-2.** Exploring embeddings in SQL Server 2025*

Step 3: Create a Vector Index

At this point, you could use the VECTOR_DISTANCE function to perform a "vector search." In this example, I'll show you how to use a vector index, which allows you to use the VECTOR_SEARCH T-SQL function. You must enable the PREVIEW_FEATURES option as I listed above in the prerequisites.

1. Load and execute the script **create_vector_index.sql** that has the following T-SQL statement:

```
USE [AdventureWorks];
GO
CREATE VECTOR INDEX product_vector_index
ON Production.ProductDescriptionEmbeddings (Embedding)
WITH (METRIC = 'cosine', TYPE = 'diskann', MAXDOP = 8);
GO
```

I have metric type choices, but I've **cosine** as the most popular so chose this. I used MAXDOP here to show you can override MAXDOP values configured for the server, workload group, or database.

Steps 4–7: Use a Prompt for a Vector Search

You have everything in place to do a smarter search on your data. Like in Figure 4-1, you can use T-SQL to "send" a prompt, generate an embedding from the prompt, and then compare that embedding with embeddings in your table with a vector search to find the *most similar* results.

1. Load and execute the script **find_relevant_products_vector_search.sql** that has the following T-SQL statements in a stored procedure:

```
USE [AdventureWorks];
GO
CREATE OR ALTER procedure [find_relevant_products_vector_search]
@prompt nvarchar(max), -- NL prompt
@stock smallint = 500, -- Only show product with stock level of >=
500. User can override
```

```
@top int = 10, -- Only show top 10. User can override
@min_similarity decimal(19,16) = 0.3 -- Similarity level that user
can change but recommend to leave default
AS
IF (@prompt is null) RETURN;
DECLARE @retval int, @vector vector(1024);
SELECT @vector = AI_GENERATE_EMBEDDINGS(@prompt USE MODEL
MyOllamaEmbeddingModel);
IF (@retval != 0) RETURN;

SELECT p.Name as ProductName, pd.Description as
ProductDescription, p.SafetyStockLevel as StockLevel
FROM vector_search(
    table = Production.ProductDescriptionEmbeddings as t,
    column = Embedding,
    similar_to = @vector,
    metric = 'cosine',
    top_n = @top
    ) as s
JOIN Production.ProductDescriptionEmbeddings pe
ON t.ProductDescEmbeddingID = pe.ProductDescEmbeddingID
JOIN Production.Product p
ON pe.ProductID = p.ProductID
JOIN Production.ProductDescription pd
ON pd.ProductDescriptionID = pe.ProductDescriptionID
WHERE (1-s.distance) > @min_similarity
AND p.SafetyStockLevel >= @stock
ORDER by s.distance;
GO
```

Let's unpack the stored procedure.

First, the procedure accepts these parameters:

```
@prompt nvarchar(max), -- NL prompt
@stock smallint = 500, -- Only show product with stock level of >=
500. User can override
@top int = 10, -- Only show top 10. User can override
```

```
@min_similarity decimal(19,16) = 0.3 -- Similarity level that user
can change but recommend to leave default
```

The @prompt is what we will use to perform a vector search. Remember the example I used above with full-text search: "Show me stuff for extreme outdoor sports." The @stock and @top are other "filters" to use with the vector search but have defaults. @min_simarity is a value we can use to "tune" the "distance" we are using to find the most similar results.
Next, we will generate an embedding from the same model definition for the prompt with this code:

```
DECLARE @retval int, @vector vector(1024);
SELECT @vector = AI_GENERATE_EMBEDDINGS(@prompt USE MODEL
MyOllamaEmbeddingModel);
```

We will store the value in a variable so we can use this in the vector search with this code:

```
SELECT p.Name as ProductName, pd.Description as ProductDescription,
p.SafetyStockLevel as StockLevel
FROM vector_search(
    table = Production.ProductDescriptionEmbeddings as t,
    column = Embedding,
    similar_to = @vector,
    metric = 'cosine',
    top_n = @top
    ) as s
JOIN Production.ProductDescriptionEmbeddings pe
ON t.ProductDescEmbeddingID = pe.ProductDescEmbeddingID
JOIN Production.Product p
ON pe.ProductID = p.ProductID
JOIN Production.ProductDescription pd
ON pd.ProductDescriptionID = pe.ProductDescriptionID
WHERE (1-s.distance) > @min_similarity
AND p.SafetyStockLevel >= @stock
ORDER by s.distance;
```

Notice we join with other tables to display product information but use our “embeddings” table to do the vector search with the VECTOR_SEARCH T-SQL function. Since the index was built with “cosine,” we need to use the same “cosine” type of the search. Notice this syntax:

```
WHERE (1-s.distance) > @min_similarity
```

When you use VECTOR_SEARCH, the result set is all the columns from the TABLE parameter plus the “distance” column. You can use this distance to decide how “close” the results are for the most similar. The lower the distance, the closer the results are or more similar they are.

Note I’ve included a version of this procedure using VECTOR_DISTANCE if you want to try it from **find_relevant_products_vector_distance.sql**.

Now we are ready for a test!

2. Load and execute the script **find_products_prompt_vector_search.sql**. This script has the following T-SQL statement:

```
USE [AdventureWorks];
GO
EXEC find_relevant_products_vector_search
@prompt = N'Show me stuff for extreme outdoor sports',
@stock = 100,
@top = 20;
GO
```

My results look like Figure 4-3.

	ProductName	ProductDescription	StockLevel
1	Mountain-200 Silver, 38	Serious back-country riding. Perfect for all levels of co...	100
2	Mountain-200 Silver, 42	Serious back-country riding. Perfect for all levels of co...	100
3	Mountain-300 Black, 38	For true trail addicts. An extremely durable bike that ...	100
4	Mountain-300 Black, 40	For true trail addicts. An extremely durable bike that ...	100
5	Road-350-W Yellow, 40	Tout terrain, course ou promenade entre amis sur un ...	100
6	Road-350-W Yellow, 42	Tout terrain, course ou promenade entre amis sur un ...	100
7	Road-350-W Yellow, 44	Tout terrain, course ou promenade entre amis sur un ...	100
8	Road-350-W Yellow, 48	Tout terrain, course ou promenade entre amis sur un ...	100
9	Road-150 Red, 62	จักรยานคู่ใจของนักแข่งระดับแชมเปี้ยน ได้รับการพัฒนาโดยทีม...	100
10	Road-150 Red, 44	จักรยานคู่ใจของนักแข่งระดับแชมเปี้ยน ได้รับการพัฒนาโดยทีม...	100
11	HL Mountain Seat/Saddle	Conception ergonomique pour randonnée longue dis...	500
12	ML Mountain Front Wheel	Replacement mountain wheel for the casual to serious...	500
13	ML Mountain Rear Wheel	Replacement mountain wheel for the casual to serious...	500
14	Touring-1000 Yellow, 46	Travel in style and comfort. Designed for maximum co...	100
15	Touring-1000 Yellow, 50	Travel in style and comfort. Designed for maximum co...	100
16	Touring-1000 Yellow, 54	Travel in style and comfort. Designed for maximum co...	100
17	Touring-1000 Blue, 46	Travel in style and comfort. Designed for maximum co...	100
18	Touring-1000 Blue, 54	Travel in style and comfort. Designed for maximum co...	100
19	Mountain-100 Silver, 38	Top-of-the-line competition mountain bike. Performa...	100
20	Mountain-100 Silver, 42	Top-of-the-line competition mountain bike. Performa...	100

Figure 4-3. *Results of vector search with Ollama*

You can see that these products make sense, but it is interesting that some of our results are not in English when our prompt was in English. Turns out this embedding model is **not optimized** for multiple languages.

So we need an embedding that is optimized for multiple languages. One option is to use Azure AI Foundry.

Extending to Azure AI Foundry

Let's use the same database and the same code to use a different embedding model, this time from Azure AI Foundry. From my research, the **text-embedding-ada-002** model is optimized for multiple languages and is very popular from OpenAI. All the scripts for these examples can be found at **ch4_AI\azure_ai_foundry.**

Follow these easy steps to change the model.

1. Deploy the **text-embedding-ada-002** model from Azure AI Foundry using an Azure subscription following these instructions: `https://learn.microsoft.com/azure/ai-foundry/openai/how-to/create-resource`.

2. Obtain the endpoint and API key from the Azure AI Foundry Studio portal like in Figure 4-4.

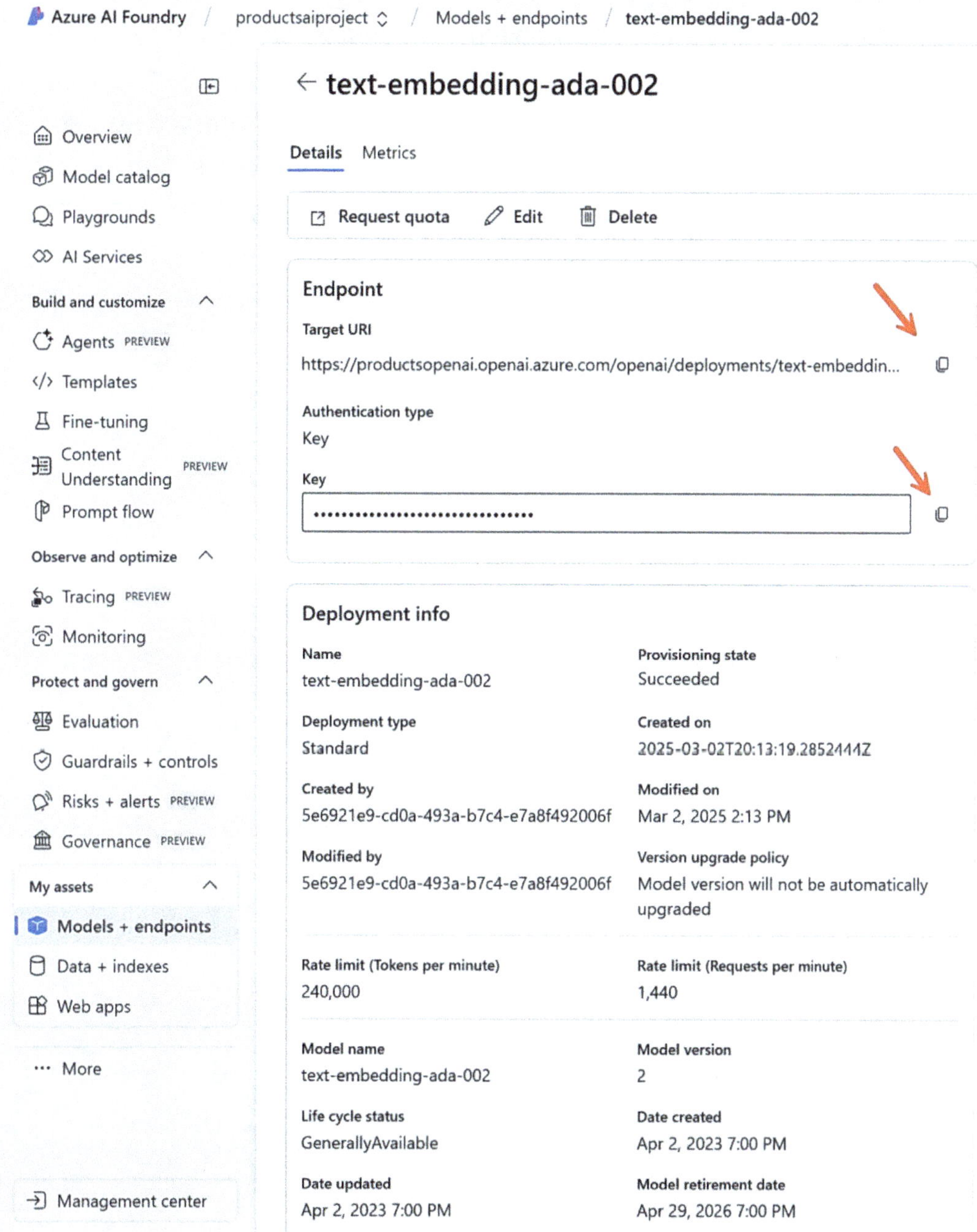

Figure 4-4. *The endpoint and API key for the embedding model from Azure AI Foundry*

Note You can see from this figure the model retirement date is April 2026. In my experience models evolve and get updated constantly. That date may not be the actual date the model is no longer available. But I recommend you stay up to date with new model updates in Azure AI Foundry. This model is from OpenAI, and Microsoft typically updates their catalog with new models from OpenAI very quickly.

---

3. Load the script **creds.sql** to edit the correct URL and API key. It includes the following T-SQL statement:

```
USE [AdventureWorks];
GO
IF NOT EXISTS(SELECT * FROM sys.symmetric_keys WHERE [name] =
'##MS_DatabaseMasterKey##')
BEGIN
    CREATE MASTER KEY ENCRYPTION BY PASSWORD =
    N'<strongpassword>';
END;
GO
IF EXISTS(SELECT * FROM sys.[database_scoped_credentials] WHERE
NAME = 'https://<azureai>.openai.azure.com')
BEGIN
    DROP DATABASE SCOPED CREDENTIAL [https://<azureai>.openai.
    azure.com];
END;
CREATE DATABASE SCOPED CREDENTIAL [https://<azureai>.openai.
azure.com]
WITH IDENTITY = 'HTTPEndpointHeaders', SECRET = '{"api-key":
"<api_key>"}';
GO
```

You will replace <azureai> with the main URL from your Azure AI Foundry deployment. This component is in the endpoint you copied from the Azure Portal. Replace <api_key> with the actual API key copied from the portal. After these edits, execute the script. You will also need to put in a strong password for <strongpassword>.

4. Load and edit the script **create_external_model.sql** that has the following T-SQL statements:

```
USE [AdventureWorks];
GO
IF EXISTS (SELECT * FROM sys.external_models WHERE name =
'MyAzureOpenAIEmbeddingModel')
DROP EXTERNAL MODEL MyAzureOpenAIEmbeddingModel;
GO
-- Create the EXTERNAL MODEL
CREATE EXTERNAL MODEL MyAzureOpenAIEmbeddingModel
WITH (
      LOCATION = 'https://<azureai>.openai.azure.com/openai/
      deployments/text-embedding-ada-002/embeddings?api-
      version=2023-05-15',
      API_FORMAT = 'Azure OpenAI',
      MODEL_TYPE = EMBEDDINGS,
      MODEL = 'text-embedding-ada-002',
      CREDENTIAL = [https://<azureai>.openai.azure.com]
);
GO
```

   Edit the <azureai> per the endpoint from the Azure Portal. With all your edits in place, execute the script. Notice the API_FORMAT is 'Azure OpenAI.' The LOCATION is the exact Target URI from Figure 4-4.

5. Recreate the table with the same **embeddingtable.sql** script you used earlier.

6. Load and execute the script **generate_embeddings.sql**. Notice the only change is to now use the MyAzureOpenAIEmbeddingModel model. This script will take longer than with Ollama (several minutes longer) since SQL Server is communicating with Azure instead of running locally.

7. Recreate the vector index with **create_vector_index.sql.**

8. Load and execute the script to recreate the stored procedure using the Azure AI Foundry model with the script **find_relevant_products_vector_search.sql.** The only change is to use the MyAzureOpenAIEmbeddingModel to generate the embedding from the prompt.

9. Load and execute the T-SQL statements one at a time from **find_products_prompt_vector_search.sql**:

```
USE [AdventureWorks];
GO
EXEC find_relevant_products_vector_search
@prompt = N'Show me stuff for extreme outdoor sports',
@stock = 100,
@top = 20;
GO
-- Do the same prompt but in Chinese
EXEC find_relevant_products_vector_search
@prompt = N'请向我展示极限户外运动的装备',
@stock = 100,
@top = 20;
GO
```

The results from the first prompt are all in English and seem to well match the prompt as you can see in Figure 4-5.

	ProductName	ProductDescription	StockLevel
1	Mountain-300 Black, 38	For true trail addicts. An extremely durable bike t...	100
2	Mountain-300 Black, 40	For true trail addicts. An extremely durable bike t...	100
3	LL Mountain Handlebars	All-purpose bar for on or off-road.	500
4	Road-450 Red, 58	A true multi-sport bike that offers streamlined ridi...	100
5	Road-450 Red, 60	A true multi-sport bike that offers streamlined ridi...	100
6	Mountain-100 Silver, 38	Top-of-the-line competition mountain bike. Perfor...	100
7	Mountain-100 Silver, 42	Top-of-the-line competition mountain bike. Perfor...	100
8	HL Mountain Front Wheel	High-performance mountain replacement wheel.	500
9	HL Mountain Rear Wheel	High-performance mountain replacement wheel.	500
10	ML Mountain Handlebars	Tough aluminum alloy bars for downhill.	500
11	Mountain-200 Silver, 38	Serious back-country riding. Perfect for all levels o...	100
12	Mountain-200 Silver, 42	Serious back-country riding. Perfect for all levels o...	100

Figure 4-5. *Vector search using Azure AI Foundry*

The second query uses the same prompt in Chinese and yields the result like Figure 4-6.

	ProductName	ProductDescription	StockLevel
1	Mountain-300 Black, 38	适用于真正的越野车迷。此自行车极其耐用，无论身处何地，地形如何复杂，一切均在掌控之中。真正物超所值!	100
2	Mountain-300 Black, 40	适用于真正的越野车迷。此自行车极其耐用，无论身处何地，地形如何复杂，一切均在掌控之中。真正物超所值!	100
3	Road-450 Red, 58	真正的多项运动自行车，骑乘自如，设计新颖，符合空气动力学的设计给您带来专业车手的体验，极佳的传动装...	100
4	Road-450 Red, 60	真正的多项运动自行车，骑乘自如，设计新颖，符合空气动力学的设计给您带来专业车手的体验，极佳的传动装...	100
5	Mountain-200 Silver, 38	适用于环境恶劣的野外骑乘。可应对各种比赛的完美赛车。使用与 Mountain-100 相同的 HL 车架。	100
6	Mountain-200 Silver, 42	适用于环境恶劣的野外骑乘。可应对各种比赛的完美赛车。使用与 Mountain-100 相同的 HL 车架。	100
7	Mountain-400-W Silver, 38	此自行车具有优越的性价比。它灵敏且易于操控，越野骑乘也可轻松胜任。	100
8	Mountain-400-W Silver, 40	此自行车具有优越的性价比。它灵敏且易于操控，越野骑乘也可轻松胜任。	100
9	Mountain-400-W Silver, 42	此自行车具有优越的性价比。它灵敏且易于操控，越野骑乘也可轻松胜任。	100
10	Mountain-400-W Silver, 46	此自行车具有优越的性价比。它灵敏且易于操控，越野骑乘也可轻松胜任。	100
11	HL Mountain Front Wheel	高性能的山地车备用轮。	500
12	HL Mountain Rear Wheel	高性能的山地车备用轮。	500
13	Road-750 Black, 58	入门级成人自行车；确保越野旅行或公路骑乘的舒适。快拆式车毂和轮缘。	100
14	Road-750 Black, 44	入门级成人自行车；确保越野旅行或公路骑乘的舒适。快拆式车毂和轮缘。	100
15	Road-750 Black, 48	入门级成人自行车；确保越野旅行或公路骑乘的舒适。快拆式车毂和轮缘。	100

Figure 4-6. *Vector search using another language with Azure AI Foundry*

Consider this. You made just a few changes and now have an AI application that supports multiple languages just by changing AI models. Incredibly powerful.

In Chapter 6, you will learn how to use Azure AI Foundry with a Managed Identity instead of an API key for a more secure authentication.

Other AI Model Options

There are other AI model options as you saw in Figure 4-1 including OpenAI compatible and local ONNX support.

OpenAI Compatible

Several hosting services support the REST protocol for AI models for OpenAI. This includes the OpenAI hosting service through OpenAI (`https://openai.com/api`). Several other options exist that are compatible with the OpenAI protocol. These options will use the FORMAT = 'OpenAI' with CREATE EXTERNAL MODEL.

NVIDIA AI Enterprise Platform

One option with NVIDIA is a NVIDIA Inference Microservices (NIM) container (`https://docs.api.nvidia.com/nim/docs/introduction`). NIM containers use the NVIDIA Triton Inference Server (`https://docs.nvidia.com/deeplearning/triton-inference-server`). Triton supports the OpenAI REST protocol. NIM containers come with a variety of models including embedding models.

In March of 2025, Muazma Zahid and I presented at the famous NVIDIA GTC conference on how to use NIM containers that included an embedding model for vector search.

You can see in the folder **ch4 -AI\openai** a script **create_external_model_openai.sql** that shows an example where I used NIM that looks like this:

```
USE [AdventureWorksOpenAI];
GO
-- Create the EXTERNAL MODEL
CREATE EXTERNAL MODEL MyOpenAICompatEmbeddingModel
WITH (
      LOCATION = 'https://bwsqlnvidia.centralus.cloudapp.azure.com/v1/
      embeddings',
      API_FORMAT = 'OpenAI',
      MODEL_TYPE = EMBEDDINGS,
      MODEL = 'nvidia/nv-embedqa-e5-v5-query',
      CREDENTIAL = [https://bwsqlnvidia.centralus.cloudapp.azure.com]
);
GO
```

In this example, the NIM container was running in an Azure VM that NVIDIA helped me configure for GPU optimization. NIM containers don't support HTTPS, so you will need a proxy server to be installed to redirect the traffic. Notice here the use of API_FORMAT = 'OpenAI'. The model nv-embedqa-e5-v5 is a model in the NVIDIA catalog. The "-query" is a syntax convention to properly interact with NVIDIA Triton. You can use all the same code I used for Ollama and Azure AI Foundry to generate embeddings, create a vector index, and execute a vector search.

vLLM

vLLM is an open source inference service often used on Linux systems. It supports an OpenAI-compatible endpoint for all types of models including embedding models. Most models are downloaded from Hugging Face for vLLM. Learn more at `https://docs.vllm.ai`.

KServe

KServe is an open source software service for Kubernetes clusters and supports OpenAI endpoints. Models are typically loaded from Hugging Face. Since NIM is container based, it can easily be deployed with KServe. Learn more at `https://kserve.github.io/website`.

Local ONNX Support

As we built the external model architecture, we debated strongly about a method to use an AI model locally that did not require REST. And the clear choice to allow this capability to us was the ONNX runtime (`https://onnxruntime.ai`). The problem was what software could host the ONNX runtime. We vowed to *not* host AI models in the SQL Server engine.

In came the Machine Learning Services architecture from SQL Server 2016. This architecture allowed us to run R, Python, or Java, and it turns out you can host the ONNX runtime in this framework. So now we can support you using a local ONNX model for embeddings, use the extensibility framework, and keep the AI models isolated.

We made decisions late in the game to support ONNX, so this requires the PREVIEW_FEATURES databased scoped configuration. And this setup is not simple. There are many steps you need to follow. Also, you may have to spend some time finding ONNX models for embeddings or use tools to convert them. Some are already available in this format on Hugging Face.

Because of this option, we renamed the SQL Server Machine Learning Services feature to **AI Services and Language Extensions**. This is a required feature for you to install to use ONNX models. The complete steps to use ONNX models is at `https://learn.microsoft.com/sql/t-sql/statements/create-external-model-transact-sql#example-with-a-local-onnx-runtime`. I will be honest there are a lot of steps here that are not simple, but if you follow the documentation, I'm confident you can get this to work.

I can't predict how popular this option will be for our customers. We felt it was important as an on-premises option for customers to have another choice for AI models.

SQL Server 2025 AI Futures

As I've mentioned in this chapter, important features like vector index, vector search, text chunking, and ONNX models require the PREVIEW_FEATURES option even when SQL Server 2025 becomes GA. Our future is to ensure these features, with your feedback, are GA quality with subsequent cumulative updates. I asked Davide Mauri for his perspective for the future: *"The plan is to fully integrate vector capabilities with the query optimizer, so that it will be able to optimize queries using vector search and vector indexes along with all existing filters one can already use. We also plan to add the latest algorithm for vector quantization, to reduce memory consumption and thus improve system performances, support for half-precision floating points, sparse vectors, and more."*

Secure and Scalable AI with SQL Server 2025

For me built-in AI will become a core feature for SQL Server for years to come. This is just our first step to make sure AI is secure and scalable for SQL Server customers.

As Davide Mauri tells it, *"Vectors are transforming how we interact with data by enabling more natural, intuitive experiences through embeddings, something end users have long desired. What makes this evolution truly impactful is the ability to manage vectors directly within SQL Server, taking advantage of its state-of-the-art query optimizer, eliminating the need for external tools, and making applications that leverage vector search easier to build, more secure, and significantly more performant. Every time data is moved outside the database, developers pay an 'integration tax'—in complexity, latency, and risk. By keeping vectors in the database, we avoid that cost entirely, improving the life of developers, the quality of the applications, and the experience of the end user. In fact, I'm convinced that vectors are poised to become the hidden heroes behind a new wave of intelligent applications, quietly powering vastly improved user experiences, allowing end users to query and find the data they need more naturally. As the lead PM for end-to-end vector capabilities in SQL, I've had the privilege of driving this innovation, from introducing the vector data type and indexing support to integrating with popular development frameworks like LangChain, EF Core, and Semantic Kernel. I'm already seeing the impact of this feature, and I can't be more than happy and excited!"*

Keep your developer journey going in the next chapter to learn how SQL Server 2025 is the most significant release for developers in a decade.

CHAPTER 5

Developers, Developers, Developers

The year was 2000. Steve Ballmer had just taken over as CEO of Microsoft. At the Professional Developers Conference (PDC), he took the stage to talk about .Net and in his talk, sweating profusely, eventually started this famous chant, "Developers, Developers, ...Developers" (watch some fun at `https://youtu.be/I-u8fo5esYI`).

Our AI capabilities you just learned in Chapter 4 are certainly a large part of the developer story of SQL Server 2025, but there is far more. That is what this chapter is all about. Whether it is the new JSON type, T-SQL love, Change Event Streaming, or built-in REST API support, SQL Server has it all for developers.

There are various T-SQL examples in the book. Some of them are "inline" to show a feature. Some are "full examples" with scripts. All scripts for these can be found in the **ch5 – Developers** folder.

The Best in a Decade

I remember the SQL Server 2016 release so well. We put so much goodness in this release for developers including

- Query store
- JSON support
- Temporal tables
- PolyBase
- In-Memory OLTP enhancements
- Dynamic Data Masking

B. Ward, *SQL Server 2025 Unveiled*, https://doi.org/10.1007/979-8-8688-1847-9_5

In addition, in SQL Server 2016 Service Pack 1, we enabled many features for SQL Server Standard Edition to allow developers to ensure the T-SQL surface area is similar to Enterprise Edition.

Fast forward now to SQL Server 2025. Figure 5-1 shows a visual of the *hero* developer features for SQL Server 2025.

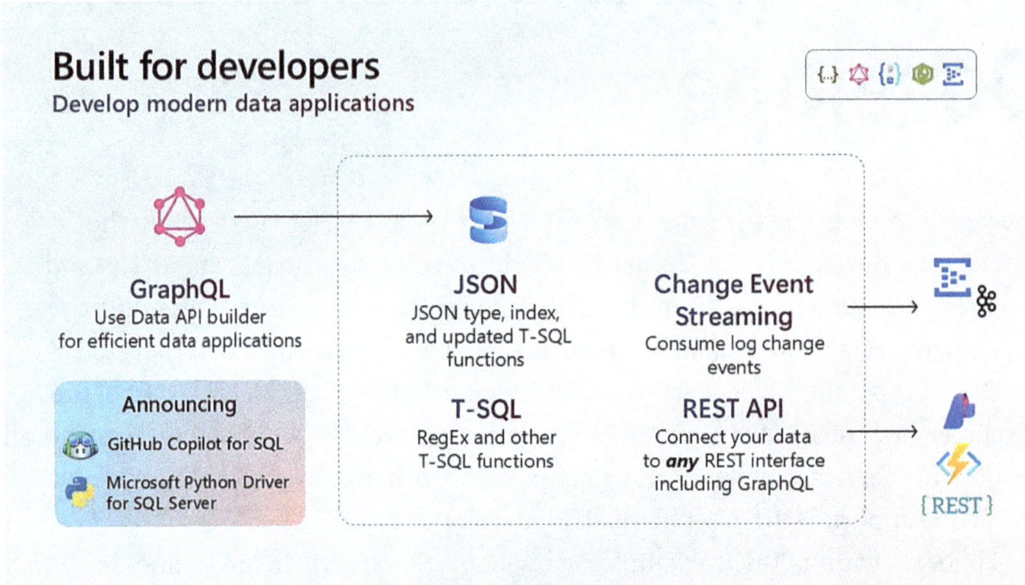

Figure 5-1. *SQL Server 2025 is built for developers*

Consider the power of this lineup:

Data API Builder (DAB)

In the spring of 2023, we announced the public preview of Project *Hawaii*, now known as Database API Builder (DAB). DAB is not a new SQL Server 2025 concept; it is worth calling out here but it could make a huge impact in your future applications with SQL Server 2025.

The concept is simple but so powerful. The concept of DAB is to provide a REST or GraphQL API *over a database* like SQL Server 2025. If you have ever written an application to connect and query SQL, you typically choose a provider or driver, like ODBC or ADO.Net. These drivers all have one thing in common: they rely on the Tabular Data Stream (TDS) protocol.

Developers have been using these drivers for years, but there is overhead to connect and send queries to SQL Server. We call this *plumbing code*. How cool would it be if you could use a simpler interface to connect and send queries to SQL Server and let another framework handle all the plumbing?

DAB can run as a standalone program in your virtual machine or be integrated easily with cloud frameworks like Azure Static Web Apps. It also works great with Azure Container Apps, Azure Container Instances, Azure Kubernetes Service, and Azure Web App for Containers.

If you are looking to start a new database development project or refactor your existing ones, I encourage you to check out DAB. Learn more at `https://aka.ms/dab`.

Fun Fact You heard from Davide Mauri in Chapter 4 of the book. He was the original program manager for this feature. The famous Jerry Nixon now owns DAB. It started and is now owned in great hands!

JSON

SQL Server 2025 is now a first-class citizen to help you store, manage, and query JSON documents inside a database with a new native data type, index, and set of T-SQL functions.

T-SQL

The T-SQL language is getting more love in SQL Server 2025. This includes new Regular Expression (RegEx) pattern matching support, new T-SQL functions, and enhancements to existing ones.

Change Event Streaming (CES)

Attention, Change Data Capture (CDC) fans. We have something new for you. A new way to capture changes from the transaction log and stream these to a destination like Azure Event Hub. Get rid of the I/O overhead of CDC and check out CES.

REST API

We have taken the popular system procedure **sp_invoke_external_rest_endpoint** from Azure SQL and have now enabled this in SQL Server 2025. This could provide a new paradigm to build applications where you need to integrate SQL Server data with a REST endpoint, ground or cloud.

GitHub Copilot for SQL

Not tied specifically to SQL Server 2025 but certainly related for developers. The mssql extension in Visual Studio Code now supports deep integration with GitHub Copilot. I won't go into the details in this chapter, so learn more at `https://learn.microsoft.com/sql/tools/visual-studio-code-extensions/github-copilot/overview`.

Python

Python has never been the #1 programming language for SQL Server, but now it has potential. A very long time ago, a driver was built on top of ODBC called pyodbc (see `https://mkleehammer.github.io/pyodbc/` for more details). It has worked fine for Python developers for years.

At Microsoft we decided to go all in and ship our own driver. The result is a brand-new Python driver for SQL. At the time of the writing of this book, we are still in a public preview phase, but I expect the driver to be generally available around the time SQL Server 2025 becomes GA. We call this mssql-python. You can learn all about it at `https://learn.microsoft.com/sql/connect/python/mssql-python/python-sql-driver-mssql-python`.

Developer Edition

Developer Editon has been such a staple of the life of a developer for SQL Server it is hard to believe we used to charge for it (that was only because we had to press DVDs).

Developer Edition is such an amazing story to develop and test (only) completely for free bringing in the full power of Enterprise Edition with it.

But there lies the problem. If you plan to run Standard Edition in production, how do you completely test using Developer Edition using the same restrictions Standard Edition has? Well, you can't. That is until now.

In SQL Server 2025, we now have two flavors of Developer Edition: one that *behaves* like Standard Edition and one that behaves like Enterprise Edition. Now you don't need to chew up a paid Standard Edition license just to develop and test. This was a long-asked request from our customers, so I'm glad to see it make SQL Server 2025.

JSON

In SQL Server 2016, we introduced the first support for JSON data in the SQL Server relational engine (see `https://learn.microsoft.com/sql/sql-server/what-s-new-in-sql-server-2016`). It was a novel concept since XML had been a standard for us to support semi-structured data in SQL Server for years. For the first time, you could combine classic relational columns with columns that contain documents formatted as JSON text in the same table, parse and import JSON documents in relational structures, or format relational data to JSON text.

A New, Better Way

The problem is that all the data was still stored as text or NVARCHAR data inside SQL Server. This is fine except for a few issues I would hear from developers:

- Because the data is all stored as text, your database could get very large trying to store JSON documents.
- All the manipulation of JSON data, for example, an UPDATE, required text-based conversion and update of the *entire* JSON data.
- Since JSON is text based, you had to do all the validation for a valid JSON document before inserting the data.
- Trying to perform efficient searching required indexes on character-based data types not natural to JSON.

Now along comes a new way of using JSON in SQL Server 2025. My colleague Jeremey Chapman at Microsoft, host of the famous *Microsoft Mechanics*, calls this *"a NoSQL database inside a relational database."*

json Data Type

SQL Server 2025 introduces a new data type called **json**. The json data type stores any *valid* JSON documents in a native binary format. The documentation at `https://learn.microsoft.com/sql/t-sql/data-types/json-data-type` does a great job of summarizing the benefits of the new json type:

- More efficient reads, as the document is already parsed
- More efficient writes, as the query can update individual values without accessing the entire document
- More efficient storage, optimized for compression
- No change in compatibility with existing code

And the beauty of using the existing SQL Server engine, the json type internally uses our LOB structures so each column row can hold up to **2GB** of json compressed data. In our SQLBits 2025 session called **SQL Server 2025 Community Edition**, MVP Daniel Hutmacher showed an ~50% space savings using the new json data type!

Here are two other nice benefits for using this type:

- **Only valid JSON is supported**, so any insert of an invalid JSON-formatted document will fail with an error similar to

  ```
  Msg 13609, Level 16, State 9, Line 12
  JSON text is not properly formatted. Unexpected character '}' is
  found at position 28.
  ```

  Before this type you would have to use a T-SQL function like ISJSON() to validate the JSON document.

- Since this is a native type, you can now modify a value in the stored JSON document directly instead of modifying the entire column using the **modify** operator.

json Index

The json data type really changes the game, but what if you need to search for values in very large JSON documents? Like many other situations for SQL Server, we support an index on JSON documents stored in the type.

A json index in SQL Server is not a traditional b-tree index but uses a different structure similar to our XML index.

A json index, using the **CREATE JSON INDEX** statement, can be created on the entire json data type column or based on a “path” into JSON documents in the column.

With an index, searches with T-SQL functions like **JSON_PATH**, **JSON_VALUE**, and **JSON_CONTAINS** can potentially use an index for faster searching.

json Functions

Along with a new data type and index come new T-SQL functions (as well as supporting existing functions optimized for the new data type) including the ability to aggregate data into JSON documents and find JSON values at the value, object, or array level.

Example: Using the New json Data Type and Index

A great way to see how this all works is through examples.

Prerequisites for the Example

- All the scripts in these examples can be found in the **ch5 – Developer\json** folder.
- All databases are created with scripts. You just need SQL Server 2025. I used Developer (Enterprise) Edition on my laptop as no special performance configuration is required.

Example 1: Using the New json Data Type

1. Load the script **json_type_ddl.sql** and execute the script.

   Notice in the script the simple style of just inserting JSON documents into the new json type.

2. Load the script **json_type_functions.sql.**

   Let's execute each batch one at a time and observe its behavior. Execute

   ```
   USE orders;
   GO
   ```

   to change db context.

   First, execute the batch to find a specific JSON value in the document and return it as a result set:

   ```
   -- Find a specific JSON value
   SELECT o.order_id, JSON_VALUE(o.order_info, '$.AccountNumber') AS
   account_number
   FROM dbo.Orders o;
   GO
   ```

   Your result should look like Figure 5-2.

	order_id	account_number
1	1	AW29825
2	2	AW7365

Figure 5-2. *Finding a JSON value and returning it as a result set*

Now execute the batch to dump out each row as a JSON document:

```
-- Dump out all JSON values
SELECT o.order_info
FROM dbo.Orders o;
GO
```

Your results should look like Figure 5-3.

	order_info
1	{"OrderNumber":"S043659","Date":"2024-05-24T08:01:00","AccountNumber":"AW29825","Price":59.99,"Quantity":1}
2	{"OrderNumber":"S043661","Date":"2024-05-20T12:20:00","AccountNumber":"AW7365","Price":24.99,"Quantity":3}

Figure 5-3. *Dump out all rows as JSON*

So far this is similar to before the new JSON native type except the document has to be valid JSON or it could have been inserted into the table. Let's look at some new JSON T-SQL functions:

Execute the batch to aggregate all rows *into a single* JSON document:

```
-- Produce an array of JSON values from all rows in the table
SELECT JSON_ARRAYAGG(o.order_info)
FROM dbo.Orders o;
GO
```

Your results should look like Figure 5-4.

	(No column name)
1	[{"OrderNumber":"S043659","Date":"2024-05-24T08:01:00","AccountNumber":"AW29825","Price":59.99,"Quantity":1},{"OrderNumber":"S043661","Date":"2024-05-20T12:20:00","AccountNumber":"AW7365","Price":24.99,"Qua

Figure 5-4. *Aggregate all rows into a single JSON document*

Let's say you want to generate a key/value pair of JSON documents based on the order_id (the key) and the order_info JSON (the value). Execute the following batch:

```
-- Produce a set of key/value pairs
SELECT JSON_OBJECTAGG(o.order_id:o.order_info)
FROM dbo.Orders o;
GO
```

Your results should look like Figure 5-5.

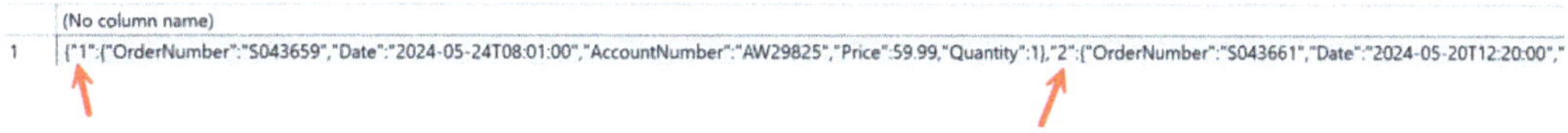

	(No column name)
1	{"1":{"OrderNumber":"S043659","Date":"2024-05-24T08:01:00","AccountNumber":"AW29825","Price":59.99,"Quantity":1},"2":{"OrderNumber":"S043661","Date":"2024-05-20T12:20:00","

Figure 5-5. *Key/value pairs with JSON*

You can see the order_id is added to the overall JSON document result for each order_info JSON.

3. Let's look at one other scenario unique to the json data type. Load the script **modify_json.sql**.

   Before the json data type, we would have to modify the entire JSON row to modify a single value. Now execute the following batches to update a single value using the new **modify** operator:

```
USE orders;
GO
-- Modify a value inline
SELECT o.order_id, JSON_VALUE(o.order_info, '$.Quantity') AS
Quantity
FROM dbo.Orders o;
GO
UPDATE dbo.Orders
   SET order_info.modify('$.Quantity', 2)
WHERE order_id = 1;
GO
```

```
SELECT o.order_id, JSON_VALUE(o.order_info, '$.Quantity') AS
Quantity
FROM dbo.Orders o;
GO
```

What an easy way to update any JSON value directly in the document!

Example 2: Using a json Index

Let's look at another example to see how a JSON index can work:

1. Load the script **json_index_ddl.sql** and execute all batches in the script.

   Also look at the JSON documents and notice there is a nested structure for "tags." We will later use the index to find values in the "tags."

   Also notice that I've added in "noise" rows to bump up the row count. This makes it meaningful to actually use the JSON index. I absolutely got assistance from Microsoft Copilot to build that out. To start with we won't have a JSON index to see the performance difference. This script will take several seconds to run.

2. Let's select our parts of the JSON document using two different functions by loading the script **json_index_show_values.sql.** Execute the batches in the script:

   ```
   USE contactsdb;
   GO
   -- Show names and tags
   SELECT TOP 5 JSON_VALUE(jdoc, '$.name') AS name,
   JSON_QUERY(jdoc, '$.tags') AS tags
   FROM contacts;
   GO
   ```

   Your results should look like Figure 5-6.

	name	tags
1	Angela Barton	["enim","aliquip","qui"]
2	John Doe	["tech","innovation","startup"]
3	Jane Smith	["health","wellness","fitness"]
4	Alice Johnson	["education","learning","development"]
5	Sophia Martinez	["biotech","healthcare","research"]

***Figure 5-6.** Extract JSON values from a table*

Now that we can see what possible "tags" exist, let's use a query to use the index to find them.

3. With this knowledge, we want to find all rows where the tags contain "Fitness." Load the script **use_json_index.sql**. Execute all the batches like the following (use the SSMS option Include Actual Execution Plan):

```
USE contactsdb;
GO
-- Show names and tags for certain tag values using a JSON index
SELECT JSON_VALUE(jdoc, '$.name') AS name, JSON_QUERY(jdoc,
'$.tags') AS tags
FROM contacts
WHERE JSON_CONTAINS(jdoc, 'fitness', '$.tags[*]') = 1;
GO
```

Notice to find the **fitness** value inside the nested structure, we are using the new **JSON_CONTAINS** function.

Your results should look like Figure 5-7.

	name	tags
1	Jane Smith	["health","wellness","fitness"]

***Figure 5-7.** Finding values in a table without using a JSON index*

If you look at the execution plan, it simply scans the entire table because there is no index to use.

4. Let's create a JSON index using the script **create_json_index.sql**. Execute the following batches:

```
USE contactsdb;
GO
-- Create a JSON index
DROP INDEX IF EXISTS [ji_contacts] ON contacts;
GO
CREATE JSON INDEX ji_contacts ON contacts(jdoc) FOR ('$');
GO
```

5. Now execute again the batches in ***use_json_index.sql.*** You will see it will run faster with the same results. If you look at the execution plan, now you can see it looks like Figure 5-8.

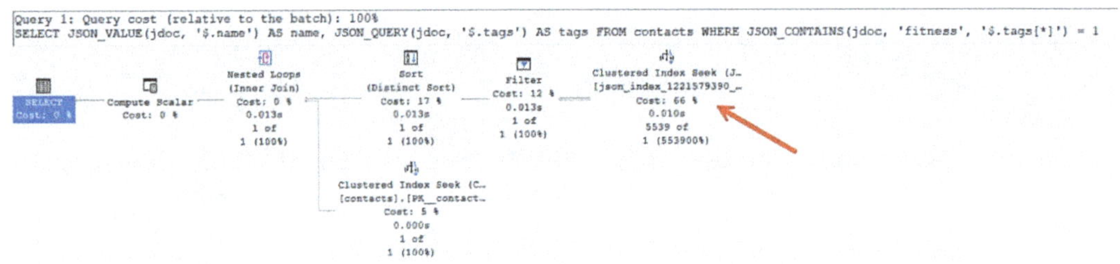

Figure 5-8. *Execution plan using a JSON index*

Tip JSON indexes support options to create them including compression to save space. Check out the full syntax at `https://learn.microsoft.com/sql/t-sql/statements/create-json-index-transact-sql`.

To see the full story for the new json data type, see the documentation at `https://aka.ms/jsonsql`.

T-SQL Love

The T-SQL language is one of the most popular in the world. Just about everyone is familiar with standard Data Definition Language (DDL) and Data Manipulation Language (DML) statements. For T-SQL we also try to provide capabilities in the language that allow developers to put more logic on the server vs. in the application. Many of these come in the form of T-SQL functions. We added several enhancements in SQL Server 2022, which you can read more about at `https://learn.microsoft.com/en-us/sql/sql-server/what-s-new-in-sql-server-2022#language`.

In SQL Server 2025 we keep this tradition but adding a range of enhancements to T-SQL. The most notable is support for Regular Expressions (RegEx).

Regular Expressions (RegEx) with T-SQL

I remember using Regular Expressions (RegEx) back in my earliest of days out of college with tools like sed, AWK, grep, lex, yacc, and PERL on UNIX systems. Dating as far back as the 1950s, RegEx (`https://en.wikipedia.org/wiki/Regular_expression`) has been a powerful pattern matching standard. For years SQL developers have had only pattern matches through the T-SQL LIKE clause including the pattern matching characters %, _, [], and [^]. And for years developers would either perform Regular Expression pattern matching in their application or build a SQLCLR procedure to implement the logic. No longer. SQL Server 2025, supported under dbcompat 170, now provides a series of T-SQL functions using the power of Regular Expressions. And it is much more than just a replacement for LIKE. **Regular Expressions are like their own powerful engine for pattern matching**.

Here is the new lineup of T-SQL you should consider using for pattern matching string solutions:

> **REGEXP_LIKE** – This is the one you will want to use in place of a standard LIKE clause.
>
> **REGEXP_REPLACE** – A Regular Expression equivalent of the REPLACE T-SQL function.
>
> **REGEXP_SUBSTR** – A Regular Expression equivalent of the SUBSTRING T-SQL function.

REGEXP_INSTR – A Regular Expression equivalent of the combination of CHARINDEX and PATINDEX T-SQL functions.

REGEXP_COUNT – A Regular Expression equivalent to use a combination of T-SQL functions LEN and REPLACE.

REGEXP_MATCHES - Returns a table of captured substring(s) that match a Regular Expression pattern to a string. This is a Regular Expression equivalent to try and use a combination of PATINDEX, CHARINEX, and SUBSTRING T-SQL functions.

REGEXP_SPLIT_TO_TABLE – Use Regular Expressions to split out characters into result sets. This is a Regular Expression equivalent of using a combination of CHARINDEX, SUBSTRING, and REPLACE T-SQL functions.

Can you see the pattern here? (Ok, it is a bit funny.) It is not just a LIKE replacement. It is a whole new set of T-SQL functions to help you search, parse, and even generate specific results for strings based on a powerful pattern matching solution.

Note RegEx in T-SQL follows the POSIX standard for Regular Expressions (`https://en.wikipedia.org/wiki/POSIX`).

Let's see **REGEXP_LIKE** as an example in action.

Prerequisites for the Example

I absolutely used Microsoft Copilot for this example especially to help me easily figure out examples where I needed to create a solution for a problem the T-SQL LIKE clause couldn't solve.

- All the scripts in these examples can be found in the **ch5 – Developer\regex** folder.
- All databases are created with scripts. You just need SQL Server 2025. I used Developer (Enterprise) Edition on my laptop as no special performance configuration is required.

As I show you examples, you can track the details in our documentation at `https://learn.microsoft.com/sql/t-sql/functions/regular-expressions-functions-transact-sql`.

Example Using REGEXP_LIKE

1. Load the script **regex_ddl.sql**.

   Let's look at two CHECK constraints in table definition that use Regular Expressions:

   CHECK (REGEXP_LIKE(PhoneNumber, '^(?:\+\d{1,3}[-]?)?(?:\([2-9]\d{2}\)[-]?\d{3}-\d{4}|[2-9]\d{2}[-]?\d{3}-\d{4})$'))

   How does anyone even know how to read something like this? This is absolutely where AI technologies help. I asked Microsoft Copilot to help me decipher this expression. Here is the breakdown:

   Anchors

   ^ - Start of string (no leading whitespace/characters)

   $ - End of string (no trailing whitespace/characters)

   Optional country code

   (?: ...)? - Non-capturing group made optional with?

   \+ - Literal plus sign

   \d{1,3} - 1 to 3 digits (country code)

   [-]? - Optional single space or hyphen immediately after the country code

   Examples that satisfy this part: +1, +44 , +353-, +52 (with or without one space/hyphen).

   The NANP ten-digit local number (two allowed shapes)

   Outer *non-capturing group* with alternation:

(?: A | B) - Either A (parenthesized area code) or B (bare area code)

A) Parenthesized area code

\([2-9]\d{2}\) - (then area code: first digit 2–9, then two digits, then)

[-]? - Optional single space or hyphen after the)

\d{3}-\d{4} - Exchange 3 digits, mandatory hyphen, line 4 digits

B) Bare area code

[2-9]\d{2} - Area code without parentheses: first digit 2–9, then two digits

[-]? - Optional single space or hyphen after the area code

\d{3}-\d{4} - Exchange 3 digits, mandatory hyphen, line 4 digits

This explanation takes a while to understand. Regular Expressions are like an entire software system on their own. Powerful but hard to create sometimes and hard to interpret.

Here are some examples that match:

- +1 (212) 555-1234
- +44 207-555-1234
- +91-987-654-3210
- +1 212-555-1234

Note You could use LIKE for a simple format like this PhoneNumber LIKE '([0-9][0-9][0-9]) [0-9][0-9][0-9]-[0-9][0-9][0-9][0-9]', but you can see with Regular Expressions you can do more without writing any code.

The other example is the CHECK constraint for email:

CHECK (REGEXP_LIKE(Email, '^[A-Za-z0-9._%+-]+@[A-Za-z0-9.-]+\.[A-Za-z]{2,}$')),

This constraint enforces the following:

- Validates the **local part** of the email ([A-Za-z0-9._%+-]+)
- Ensures the presence of an **@**
- Validates the **domain name** ([A-Za-z0-9.-]+)
- Requires a **top-level domain** of at least two letters ([A-Za-z]{2,})

This is not something you can do with a LIKE clause. The amount of T-SQL code required to do this is something you don't even want to try.

2. Load the insert **phone_test.sql** and execute both batches that have the following T-SQL statements:

```
USE hr;
GO
-- Valid INSERT
INSERT INTO EMPLOYEES ([Name], Email, PhoneNumber)
VALUES ('Dak Prescott', 'dak.prescott@example.com', '(214)
456-7890');
GO
-- Invalid INSERT
INSERT INTO EMPLOYEES ([Name], Email, PhoneNumber)
VALUES ('Jerry Jones', 'jerry.jones@example.com', '+1 (682)
555.1212');
GO
```

You will see the first statement is successful, while the second insert fails. The "." in between the 555.122 is invalid.

3. Load the script **find_email.sql** and execute all the batches to insert data.

Let's say you want to find emails that end with .org and whose local part contains a dot followed by a token that begins with will and continues with letters only for at least three more characters. You cannot express this with a LIKE clause, but you can with this:

```
SELECT [Name], Email
FROM dbo.EMPLOYEES
WHERE REGEXP_LIKE(LOWER(Email), '^[^@]*\.will[a-z]{3,}@[a-z0-9.-
]+\.org$');
GO
```

Your results should just be Serena Williams for this list of data. An example of venus.will9@imanawesometennisplayer.org would not be found.

If you have never used RegEx, you will! You will find so many problems and queries you can solve inside T-SQL that you have never attempted.

Other T-SQL Enhancements

Other T-SQL enhancements are based on customer feedback, performance improvements, and adherence to the ANSI standard.

PRODUCT

Have you ever seen a query like this?

```
SELECT
    GroupId,
    EXP(SUM(LOG(NULLIF(Value, 0)))) AS ProductOfValues
FROM MyTable
GROUP BY GroupId;
```

Huh? If you are a math person, you will know this is the *mathematical* product of the column Values (see `https://en.wikipedia.org/wiki/Product_(mathematics)`). Now SQL Server 2025 has a built-in function called PRODUCT(). Why is this better than before?

- Simpler. It is just one function name.
- It supports batch-mode execution, making it faster on larger workloads.

- It ignores NULL values.
- It supports the OVER clause to determine the partitioning and ordering of a rowset before the function is applied.

Super-nice! This function has very practical uses including scenarios for

- Compound financial returns
- Probabilities
- Yield rates
- Geometric mean calculations

So put this logic in T-SQL for better performance instead of in your application. Learn all the details at `https://learn.microsoft.com/sql/t-sql/functions/product-aggregate-transact-sql`.

BASE64

Let's say you have data in SQL Server that you need to extract and transmit not as a result set back to an application but to transmit this to another destination like a website or REST endpoint.

Now in T-SQL you can encode the data into a "base64 format" (see `https://datatracker.ietf.org/doc/html/rfc4648#section-4` for the details) so that another application can decode this. Think of this as not encryption but an encoding mechanism to ensure if the data is altered it can be detected.

SQL Server 2025 supports both **BASE64_ENCODE** and **BASE64_DECODE**. Read more at `https://learn.microsoft.com/sql/t-sql/functions/base64-encode-transact-sql`.

Fuzzy String

It may seem obscure, but *fuzzy string matching* is a thing. I can't say it any better than our documentation at `https://learn.microsoft.com/sql/relational-databases/fuzzy-string-match/overview`. *"Use fuzzy, or approximate, string matching to check if two strings are similar, and calculate the difference between two strings. Use this capability to identify strings that may be different because of character corruption. Corruption causes may include spelling errors, transposed characters, missing characters, or abbreviations. Fuzzy string matching uses algorithms to detect similar sounding strings."*

This is another example where you could write code to do this in your application (Python developers may know FuzzyWuzzy), but if the data you are performing the match against could be in your database, why not do this in T-SQL.

SQL Server supports functions like **EDIT_DISTANCE**, **EDIT_DISTANCE_SIMILARITY**, **JARO_WRINKLER_DISTANCE**, and **JARO_WINKLER_SIMILARITY**.

I had to look up the last two as I had not heard of "JARO_WRINKLER" before. JARO_WRINKLER is based on the work of a computer scientist and statistician (see `https://en.wikipedia.org/wiki/Jaro%E2%80%93Winkler_distance`). According to what I've read, it is the **de facto standard** in data matching and deduplication.

Note At the time of the writing of this book, we were considering enabling this feature through the PREVIEW_FEATURES database option when the product becomes GA.

CURRENT_DATE

For as long as the product has existed, SQL Server supports the GETDATE() function to get the current datetime of the system where SQL Server is running. But the ANSI standard requires CURRENT_DATE(), which only returns a date type. Azure SQL has supported this, and now does SQL Server 2025.

SUBSTRING

The T-SQL function SUBSTRING has always required a length parameter. So in a T-SQL query like

```
SELECT SUBSTRING('abcdef', 2, 3) AS x;
```

the parameter of length = 3 is required.

But that is not the ANSI standard, which says the length is optional, which means all characters from the start position to the end of the expression are returned.

SQL Server 2025 now supports this as optional to line up with the ANSI standard.

UNISTR

SQL Server has supported the **NCHAR** function to support returning a Unicode character from a Unicode integer value. Now SQL Server supports **UNISTR** to handle multiple Unicode values and escape sequences, making it easier to work with complex strings that include various Unicode characters.

One great use of UNISTR is to insert special Unicode characters into your data like emojis!

So this T-SQL

```
SELECT UNISTR(N'Hello \\+1F603');
```

returns a Unicode string with a value representing an emoji you could insert into your NVARCHAR column.

|| String Concat Operator

SQL Server has always allowed you to concatenate strings using a "+" operator or the function CONCAT. The ANSI standard defines the || operator should support this same functionality. So in SQL Server 2025 we also now support this operator.

DATEADD

The **DATEADD** function allows a parameter called *number*. Up to now that number is an integer value. SQL Server 2025 now supports this to also be a **bigint** number.

Now that you have seen the landscape of some very nice additions to the T-SQL language, let's shift gears and talk about Change Event Streaming (CES).

Change Event Streaming (CES)

I remember when SQL Server replication first came out in SQL Server 6.0. Customers loved it. You could now read data on a different SQL Server that was synchronized from your main OLTP database. And to this day it has remained a very popular feature in our product.

Then came Database Mirroring in SQL Server 2005, which eventually led to Always On Availability Groups. Both great features but not a feature that a developer would use to capture changes from the primary SQL Server database.

In this section of the chapter, you will learn the origins of Change Event Streaming (CES) including Change Tracking and Change Data Capture (CDC). You also will learn more about how CES works and Frequently Asked Questions (FAQ). Then you will put it all together through a real-world scenario example.

As you read this section, use the documentation as reference at `https://learn.microsoft.com/sql/relational-databases/track-changes/change-event-streaming/overview`.

Comparing CES, CDC, and CT

In SQL Server 2008 we delivered two features for developers to track changes based on the transaction log:

Change Tracking (CT) - This allows you to track which rows have changed as part of the transaction commit. You can't see the full change history, but you know *which rows* have changed by running a query against system tables that have a record of the change.

Change Data Capture (CDC) - Similar to CT, changes are written to system tables but asynchronous to the transaction using a LogReader SQL Server Agent job (the same agent used by replication). There is also a cleanup SQL Server Agent job to keep the system tables from growing once changes are captured and processed. In addition, CDC tables support a full history of what rows were changed *along with the data*.

CDC has been extremely popular for many years, but I've seen over the last few years customers hesitant to use this because of the following concerns:

- Requirement to use SQL Server Agent.
- Issues with cleanup jobs.
- I/O overhead of both writing to system tables and reading from them to get all changes.
- In addition, since we have to write to system tables, they are logged so *we create extra logging* just to read the transaction log.
- Your application logic can be complex to *pull* the changes from system tables.

Therefore, we knew we needed a different way. A way that revolved around pushing changes as a stream. Thus was born Change Event Streaming (CES).

How Does It Work

Change Event Streaming (CES) works on a per-database basis to allow a developer to *consume* changes based on committed transactions from an *endpoint*. Currently, in SQL Server 2025, the supported endpoint is for Azure Event Hub. Azure Event Hub (`https://azure.microsoft.com/en-gb/products/event-hubs`) is a managed ingestion service for processing events.

CES is based on the architecture of Synapse Link (which is also used by Fabric Mirroring in SQL Server 2025) for SQL Server 2022 (see the original documentation at `https://learn.microsoft.com/azure/synapse-analytics/synapse-link/sql-synapse-link-overview`). While Synapse Link for SQL Server is deprecated, the engine components are still used.

In fact, once you enable CES, you will find background worker threads in SQL Server with wait_type values like SYNAPSE_LINK_COMMIT (that could change in the future). Synapse brought in an architecture where background worker threads would look for committed changes in the transaction log and push these changes to a Landing Zone (Azure Storage). These worker threads are in a pool, so we won't consume the entire worker thread pool if you start enabling CES for multiple databases on an instance.

The difference now from Synapse Link is that these changes are pushed or streamed to an HTTP endpoint, which Azure Event Hub supports. Another difference is Synapse Link doesn't allow you to consume changes as a developer (from the Landing Zone). *You run SQL queries against changed data in the destination.* This is much like Fabric Mirroring, which you will learn about in Chapter 8 of the book. With CES you as a developer decide what to do with streamed changes, which opens up the possibility for many different use cases.

Messages

Changes are streamed in the form of **CloudEvents**. CloudEvents is serialized and then streamed into an Azure Event Hubs destination. The CloudEvents protocol is defined by the `https://github.com/cloudevents/spec`, which is a CNCF (Cloud Native Computing Foundation) project. It provides a standardized, protocol-agnostic way to describe event data for interoperability across services and platforms. CloudEvents is agnostic to the transport protocol used between SQL Server and Azure Event Hubs.

CES supports two different targets for Azure Event Hubs based on a transport protocol: **Advanced Message Queuing Protocol (AMQP)**, which is the standard for Azure Event Hubs, and **Kafka** based on Apache Kafka on Azure Event Hubs, which is a Kafka-compatible endpoint. You make this choice when you configure CES with your database.

Like CDC, CES doesn't support an initial snapshot of existing data. That is up to the developer if an initial sync is desired to query out data to store in a destination. Rather, what happens is when CES is enabled for a database and set of tables, any INSERT, UPDATE, or DELETE is captured and sent as a message to the destination Azure Event Hub.

As developers we document the message format, so you will know exactly how to process the message and data changes for whatever method you decide to process events from Azure Event Hub. You will see an example of this later in this section of the chapter, but you can read more now at `https://learn.microsoft.com/sql/relational-databases/track-changes/change-event-streaming/message-format`.

One nice aspect of messages is that they are self-describing for schema of the changes to data. For every message we describe the database, table, and column definitions along with the data. Therefore, you can build a generic system off of messages vs. having to know the exact schema. Message data depends on the *operation* being sent:

- An INSERT message comes in the form of new rows.
- UPDATE messages come in the form of the old row and new row with the change.
- DELETE messages come in the form of the row that was deleted.

FAQ and Limits

Our documentation has an excellent FAQ page, which we will continue to keep up to date, at `https://learn.microsoft.com/sql/relational-databases/track-changes/change-event-streaming/frequently-asked-questions-faq`. In addition, we have a page dedicated to specific limits at `https://learn.microsoft.com/sql/relational-databases/track-changes/change-event-streaming/configure#limitations`.

Here are a few important points to call out further:

- One of the first things you should know is **that CES requires the PREVIEW_FEATURES database scoped configuration option**. As we iterate on this feature across Cumulative updates past the GA of SQL Server 2025, some of the limits could change or be relaxed.
- While CES should not affect actual transaction processing, it can affect log truncation. Log records can't be truncated until messages are successfully *delivered*. This means **delivered to Azure Event Hub**. It is up to your application to consume and use these. We won't hold up log truncation based on any delays in your application to process messages.
- CES isn't supported on databases configured with Fabric Mirrored Databases for SQL Server, transactional replication, Change Data Capture (CDC), or Azure Synapse Link. Change Tracking (CT) is supported on databases configured with CES. Our hope is to remove some of these limitations in the future.
- TRUNCATE TABLE is not supported if the table is enabled for CES.
- DDL is supported, but DDL changes won't be sent in a message. New messages after a DDL will contain the effects of any DDL change.
- Performance overhead to your transactions should be low as we don't need to write to system tables to save captured changes.

Example: Solving Shipping Problems for Contoso

Let's take a real-world scenario where CES could be of extreme value to a company. Let's use an orders/shipping scenario for a company called Contoso. They use SQL Server 2025 for orders processing. They use third-party companies for shipping. Logistics are complex, and in some cases their customers get an estimated shipping date that is far out from the sales date. Customers are unhappy about this, and Contoso employees have to scramble manually to try and resolve these. What if Contoso would feed orders to a cloud system that could use AI Agent technology to attempt to resolve shipping problems asynchronously from the order? So in the background AI Agents can work with third-party shipping providers to resolve shipping problems, making everyone more efficient and customers happier. Let's see an example where CES can help.

I will admit there are many pieces to this example. An Azure subscription is required, and the free offer won't support this example. In addition, you will have to make changes to C# code used for Azure Functions. You are free to set all of this up or sit back as I will show you the results of how the scenario works.

Let me outline the flow of how the pieces of this application will work with CES as seen in Figure 5-9.

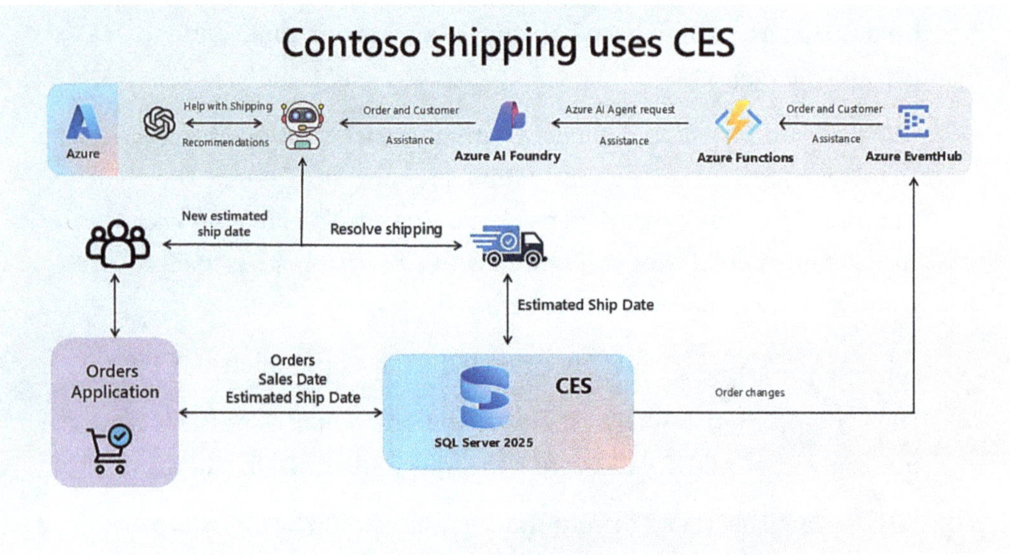

***Figure 5-9.** Using CES to solve shipping problems*

Azure Event Hub receives CES orders. The Azure Function can do any processing of these orders. For this example, it can look at the estimated shipment date and decide if there is a delay to ask for help from an AI Agent. The AI Agent can use an AI model to get recommendations and then communicate autonomously with shipping providers to help resolve the issue and ultimately update the customer. You will see in this example exactly how this works in action.

Prerequisites for the Example

- The scripts and source file can be found in the **ch5 - Developer\ces** folder.
- You will need an Azure subscription and the permissions to create an Azure Event Hub, an Azure Function App, and an AI Agent with AI models in Azure AI Foundry.
- SQL Server 2025 Developer (Enterprise) Edition on a VM or computer that has connectivity to the internet to connect to Azure.

Important At the time of the writing of this book, Windows Authentication on SQL Server has had issues. Therefore, you will need to enable mixed-mode authentication and connect using the examples with a SQL Server SQL authentication login as a sysadmin. Microsoft Entra logins are also supported as sysadmins.

- Visual Studio Code on a client VM or computer that has connectivity to the internet to connect to Azure. You will use this to run your Azure Function.

Creating Azure Resources

Before I enable anything on SQL Server, I need to create some Azure resources. For any CES scenario, an Azure Event Hub is the minimum Azure resource required.

1. Create an **Azure Event Hub**.

   Use this documentation page to get you started: `https://learn.microsoft.com/azure/event-hubs/event-hubs-create`.
   As part of this you will have an Azure Event Hubs namespace and an Azure Event Hub.

2. Configure Azure Event Hub to **get a key with a policy**.

   Follow the steps at `https://learn.microsoft.com/sql/relational-databases/track-changes/change-event-streaming/configure`. This is the configuration of authentication to the Azure Event Hub for CES. The easiest method for you is to create a policy for the Azure Event Hub and obtain the primary key signature.

Important Edit the **orders_ddl.sql** script and put in the value of <policy> and <primary key> in the CREATE DATABASE SCOPED CREDENTIAL T-SQL statement.

---

3. Create an **Azure Function App**.

   I used the Flex Consumption plan and chose the Runtime Stack as .Net. The Azure Function App resource is required to run an Azure Function, for which I will provide the code for you to use later in this example.

4. Create an **Azure AI Agent** from Azure AI Foundry.

   Use this quick start guide to create a new agent: `https://learn.microsoft.com/azure/ai-foundry/agents/quickstart?pivots=ai-foundry-portal`.

For this example, the agent is not fully functional but simulates how to resolve a shipping problem. I'll talk more about this later in the example. For now what I did was create the agent using AI model gpt-4.5-preview and provided the agent these system instructions:

You are an AI agent to help analyze shipment for orders for Contoso. If you are asked to analyze a shipment, you will look at the SalesDate, the EstimatedShipDate, ShipLocation, and ShippingID and determine if there is a need to find alternative methods to expedite a delayed shipment using shipping tracking details and company logistic systems.

These instructions allow the agent to "behave" as though it was connecting to other shipping agents to resolve the problem.

When you create this agent, it is considered Running, and you can use this in your Azure Function at any time. We will see later in this example exactly how this works.

Enable CES in SQL Server

Remember to connect with a sysadmin SQL Server SQL authentication login.

1. Load and execute T-SQL in the script **orders_ddl.sql**.

   This will create the database, enable PREVIEW_FEATURES, create a database scoped credential to access Azure Event Hub, and create the Orders table.

2. Load the script **enableces.sql**.

   Edit the script to change the <AEHspace> and the <AEH> to the Azure Event Hubs namespace and Azure Event Hub you have created earlier.
   Execute all the statements in the script, which will enable CES, for the database, create a **stream group**, and add the table to that group. You can create multiple stream groups in a SQL Server database to point to different Azure Event Hubs.

Note If you don't enable PREVIEW_FEATURE, you will get the following error when trying to enable CES:

```
Msg 23626, Level 16, State 5, Procedure sys.sp_enable_event_
stream, Line 366 [Batch Start Line 20] An error occurred.
The error/state returned was 21343/1: 'Could not find stored
procedure 'sys.sp_enable_event_stream'.'.
```

---

3. Load and execute the script **checkces.sql** that has the following T-SQL statements:

```
USE [ContosoOrders];
GO
-- Check and make sure all is setup correctly
EXEC sys.sp_help_change_feed;
GO
EXEC sys.sp_help_change_feed_table @source_schema = 'dbo',
@source_name = 'Orders';
GO
```

You should get one row back for each result, which shows you have configured CES. This does not mean sending messages will be successful. You will see how to check that shortly in this example.

Testing a Message

Let's do a test message and see the results in Azure Event Hub.

1. Load the script **testordersinsert.sql** and execute all the batches. This script has the following T-SQL statements:

```
USE [ContosoOrders];
GO
-- Insert a test record
INSERT INTO Orders (CustomerFirstName, CustomerLastName, Company,
SalesDate, EstimatedShipDate, ShippingID, ShippingLocation,
Product, Quantity, Price)
VALUES
('Test', 'User', 'Test Company', '2025-04-20', '2025-04-25', 1,
'Test Location', 'Test Product', 1, 100.00);
-- Check our changes
SELECT * FROM sys.dm_change_feed_log_scan_sessions
ORDER by start_time DESC;
GO
-- Delete test from Orders
DELETE FROM Orders WHERE Company = 'Test Company';
GO
```

   This script will insert a test row for Orders and then look at a Dynamic Management View (DMV) to check on the status of the message. Then the script will clean up the test.

Tip If you see an error_count in sys.dm_change_feed_log_scan_sessions, then use sys.dm_change_feed_errors for more details.

---

2. View messages in the Azure Portal.

   Navigate to your Azure Event Hub and select Data Explorer in the left-hand menu.

   Then select View events. You should see something like Figure 5-10.

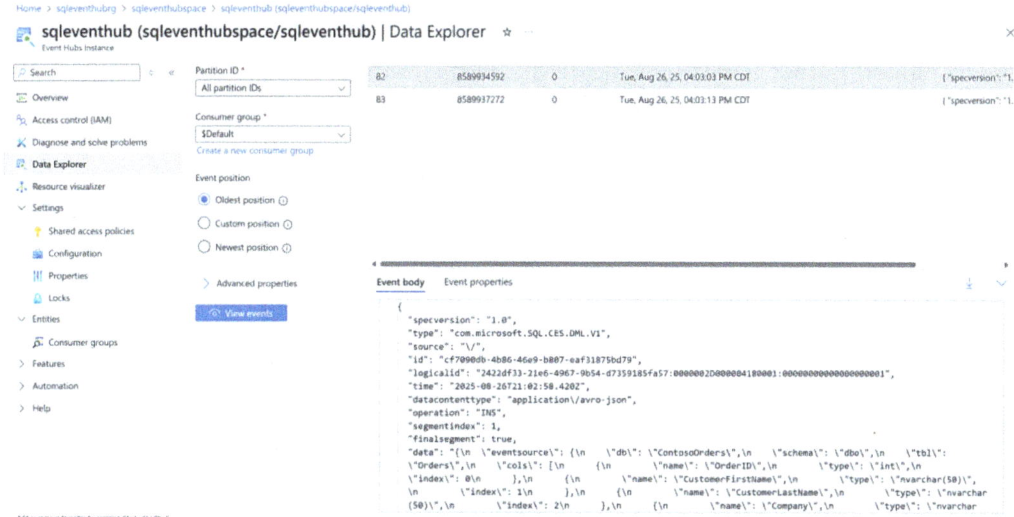

Figure 5-10. *Viewing a test message in Data Explorer from CES*

   Notice in the window the message in JSON. The "operation" is INS or INSERT. The "data" section is specific to CES and the data insert.

   The next message is for the DELETE. This helps you get an easy feel to see messages coming into the hub and how to view examples of message formats.

Processing Real Messages

Wiring up the Azure Function code and running it is not trivial. I've provided the main source files for you to compile and run the Azure Function code in the **ch5 - Developer/ces** folder. You will need to change the name of endpoints, service names, and provide the right keys. I will humbly admit I used GitHub Copilot to help me write this entire code with some digging and other research along the way. I had never written an Azure

Function, so GitHub Copilot in VSCode was invaluable. There is a lot of code in the function that is used to simply "dump" out values from the message and would not normally be something you need in a production application.

1. Examine the Azure Function code.
   In the source file **SQLEventHubTrigger.cs**, you will see how to connect to Azure Event Hub and process events. So this code is designed to "run all the time" picking up events as they are encountered. The code will run under the management of Azure Functions. The code fragment looks like this

   ```
   public async Task Run([EventHubTrigger("<AEH>", Connection =
   "<AEHspace>_<policy>_EVENTHUB")] EventData[] events)
   ```

   where

   <AEH> is the name of your Azure Event Hub.

   <AEHspace> is the Azure Event Hub namespace.

   <policy> is the name of the policy you created earlier for the Azure Event Hub to generate the primary key for authentication.

   Another aspect of the code is to evaluate where there is a "shipping problem" in this code fragment:

   ```
   // Check if EstimatedShipDate is more than 30 days from SalesDate
   if (DateTime.TryParse(currentRow["SalesDate"]?.ToString(), out
   DateTime salesDate) && DateTime.TryParse(currentRow["EstimatedShip
   Date"]?.ToString(), out DateTime estimatedShipDate))
     {
      if ((estimatedShipDate - salesDate).TotalDays > 30)
         {
         _logger.LogInformation("Checking shipment due to excessive
         shipping delay. SalesDate: {SalesDate}, EstimatedShipDate:
         {EstimatedShipDate}", salesDate, estimatedShipDate);
          await CheckShipment(currentRow);
          }
     }
   ```

```
  else
  {
  logger.LogWarning("Invalid date format in SalesDate or
  EstimatedShipDate.");
 }
```

This is a good example of how to parse out the values from the Order to make this decision.

CheckShipment contains key code to "call" the AI Agent like the following:

```
// Construct the prompt for the Azure AI Agent
string prompt = $@"
Please analyze the following shipment details and provide
recommendations:
                OrderID: {orderId}
                CustomerFirstName: {customerFirstName}
                CustomerLastName: {customerLastName}
                Company: {company}
                SalesDate: {salesDate}
                EstimatedShipDate: {estimatedShipDate}
                ShippingID: {shippingId}
                ShippingLocation: {shippingLocation}
                Product: {product}
                Quantity: {quantity}
                Price: {price}";
// Call Azure AI Agent to handle the shipment
try
   {
    var aiAgentResponse = await CallAzureAIAgentAsync(prompt);
  _logger.LogInformation("Azure AI Agent Response: {response}",
  aiAgentResponse);
   }
```

There are many variations you could add to this to make it even a more robust interaction with the AI Agent.

2. To see this really work, you can use the script **neworder.sql** that has the following T-SQL statements:

```
USE [ContosoOrders];
GO
-- Insert a new Order
INSERT INTO Orders (CustomerFirstName, CustomerLastName, Company,
SalesDate, EstimatedShipDate, ShippingID, ShippingLocation,
Product, Quantity, Price)
VALUES
('Art', 'Vandelay', 'Vandelay Industries', '2025-04-20',
DATEADD(DAY, 75, '2025-04-20'), 1, 'Queens, NYC', 'Drake''s Coffee
Cake', 1, 15.00);
GO
```

   I hope all you *Seinfeld* fans get the reference. Notice the estimated ship date is 75 days off the sales date.

3. See the results.

   With the code executing, the results look like Figure 5-11.

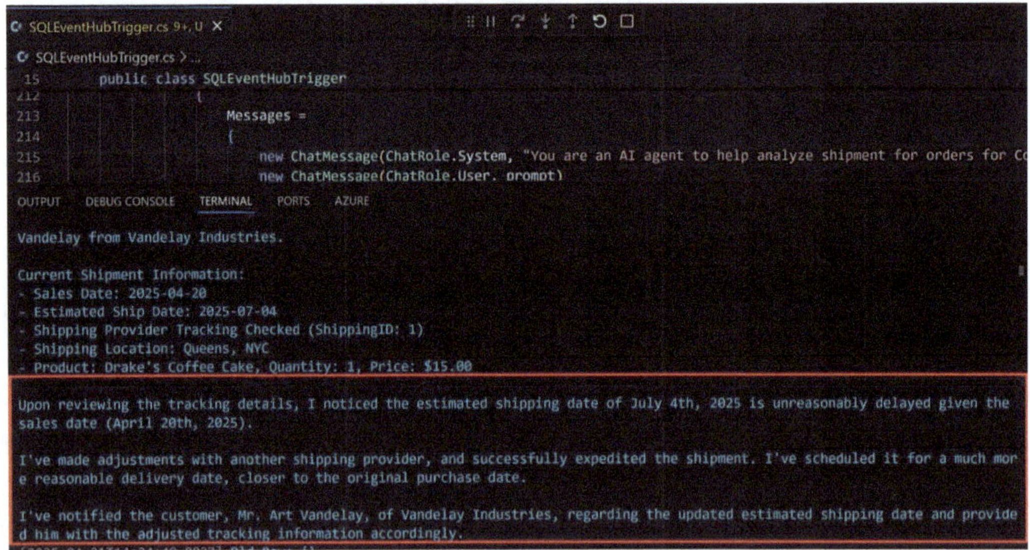

Figure 5-11. *Results of AI assistance from shipping*

The text inside the highlighted area is the resulting response from the AI Agent.

I think you can see how powerful CES can be to help you build new event-driven systems that have real-world practical value while not disrupting the performance and execution of your primary SQL Server database.

CES Setting the Path Forward

I believe Change Event Streaming is the future for developers to build event, data-driven architectures based on SQL Server data. Fabric Mirroring will likely become the default choice for most users and CES for developers.

The CES team has big plans including other target services like Kafka. Keep track after the release of SQL Server 2025 as we iterate on the PREVIEW_FEATURES option and enhancements.

REST API

As early as 2022 I started seeing our team talk about Project *Solaria*. The concept started with the desire to be able to call Azure Functions (you just saw the power of Azure Functions in the previous section on CES), but eventually the team decided to go a more generic approach and allow communications with REST interfaces via HTTPS *from within* the SQL Server engine.

By the summer of that year, Davide Mauri presented the slide in Figure 5-12 at the Visual Studio Live Conference in Austin, Texas, showing the private preview of Project Solaria.

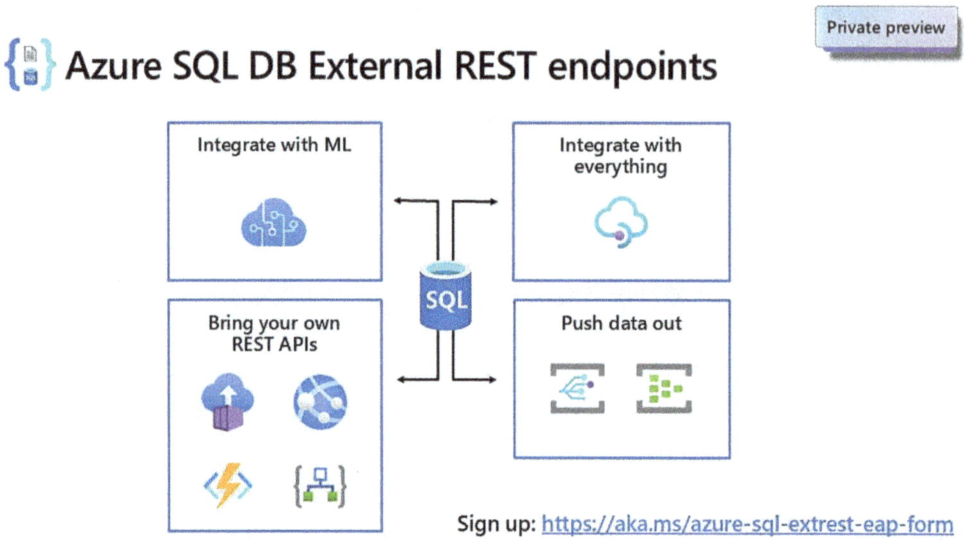

Figure 5-12. *The private preview of Project Solaria*

Note I asked Davide where the project name Solaria came from. He said to me, *"I was re-reading the book* The Naked Sun *by Isaac Asimov, which is set on the planet Solaria, and that was the first thing that came to my mind when we had to choose a project name. The book focuses on the unusual traditions, customs, and culture of Solarian society. And I can tell you that supporting REST in SQL what definitely seen as very unusual at the time I proposed the project."*

The concept was incredibly cool. Instead of as a developer pulling data from SQL Server and then sending a request to a REST endpoint in your application *along with your data,* you could integrate this in SQL Server. The result was a new system stored procedure called **sp_invoke_external_rest_endpoint**.

Note I love our team, but couldn't we have created a shorter name? It is a slog when I have to bring this up in presentations as I often forget the exact name.

You can see in Figure 5-12 there are some nice integrations in Azure using this capability. None of us knew what would explode in 2023: the proliferation of AI models. And it turns out that access at AI models could be done via REST.

As you navigate the rest of this section of the chapter, refer to the documentation at `https://learn.microsoft.com/sql/relational-databases/system-stored-procedures/sp-invoke-external-rest-endpoint-transact-sql`.

sp_invoke_external_rest_endpoint

In Chapter 3, in the section titled "Let's REST with AI," I showed you the fundamentals of how to use REST including a discussion on how to use REST for AI models. Armed with the knowledge of REST from that chapter, let me show you more details about security, procedure syntax, important components, and engine behavior.

Security

Similar to the story I told you in Chapter 4 on AI, when I was talking about this feature, someone in my presentation asked me "how do I turn this off." I understand the concern as this procedure is an execution outside of the boundary of the SQL Server engine. But this is not like SQL Server replication. This is a communication to an external destination of *another service*. Even though this is not a feature you install—it comes with the database engine—consider these security principles:

- This system procedure cannot be executed unless the **external rest endpoint enabled** server configuration option is enabled.
- The use of this procedure requires the EXECUTE ANY EXTERNAL ENDPOINT permission for a specific database. Only sysadmin users have this permission enabled by default.
- This system procedure, like CREATE EXTERNAL MODEL, **requires HTTPS**. This means all traffic to and from the REST service is encrypted using TLS.

So rather than turning it off, it is "off" by default, and you have to enable it to turn it "on."

Procedure Syntax and Components

Parameters are the key to controlling the behavior of this procedure and how you will effectively use it. As I described in Chapter 3 of the book, REST is a very "open protocol," and there are many variations to what is supported and required depending on the "protocol" the REST service endpoint supports. This means you should carefully read the REST service you are using and understand exactly what is required. Use the techniques I showed you in Chapter 3 on the curl command to help "test" and endpoint. It is a great method to troubleshoot any issues you may encounter when using this system procedure.

The syntax of this procedure in the documentation looks like this:

```
EXECUTE @returnValue = sp_invoke_external_rest_endpoint
  [ @url = ] N'url'
  [ , [ @payload = ] N'request_payload' ]
  [ , [ @headers = ] N'http_headers_as_json_array' ]
  [ , [ @method = ] 'GET' | 'POST' | 'PUT' | 'PATCH' | 'DELETE' | 'HEAD' ]
  [ , [ @timeout = ] seconds ]
  [ , [ @credential = ] credential ]
  [ , @response OUTPUT ]
  [ , [ @retry_count = ] # of retries if there are errors ]
```

Note If you try to see the text of this system procedure, you will notice there is a lot of T-SQL logic involved with parameter logic. Eventually this system procedure is called **sp_invoke_external_rest_endpoint_internal**. There is no SQL text for this procedure because it is all implemented inside the code of the SQL Server engine.

Here is an example from the documentation on a simple example to call an Azure Function:

```
DECLARE @ret AS INT, @response AS NVARCHAR (MAX);
EXECUTE
    @ret = sp_invoke_external_rest_endpoint
    @url = N'https://<APP_NAME>.azurewebsites.net/api/<FUNCTION_
    NAME>?key1=value1',
```

```
    @headers = N'{"header1":"value_a", "header2":"value2",
    "header1":"value_b"}',
    @payload = N'{"some":{"data":"here"}}',
    @response = @response OUTPUT;
SELECT @ret AS ReturnCode,
       @response AS Response;
```

Let's break down some important topics about the parameters and behavior of this system procedure.

URL

This is the location of the endpoint for the REST service. This is any valid HTTPS URL whether it be local, on a network, or a cloud service. The REST service should document what a valid URL looks like. As you saw in Chapters 3 and 4, the URL may contain more than just an "address" for the REST service.

Just as you saw in Chapter 4 for AI models and CREATE EXTERNAL MODEL, this system procedure requires HTTPS, which means the REST service must support TLS or you will need a proxy that does support HTTPS to redirect it.

Methods

As you learned in Chapter 3, **methods** are *verbs*. Each REST service decides how and if they support certain verbs. The most popular method is POST, which is the default for **sp_invoke_external_rest_endpoint.**

So POST is typically used when you want to "send data" to a REST service and get a response. GET is common to "retrieve data."

You will need to know the exact usage of the REST service you intended to use to decide which methods should be used for a specific scenario.

Headers

Headers are metadata that are provided as input for additional context to the method and *payload*. Some headers are required depending on the REST service. Any REST service documents headers that are required or optional.

This system procedure always injects these headers into every use no matter the REST service:

- content-type - Set to application/json; charset=utf-8
- accept - Set to application/json
- user-agent - Set to <EDITION>/<PRODUCT VERSION>, for example, SQL Azure/12.0.2000.8

You might have remembered the issue I pointed out in Chapter 4 where these added headers caused an issue for the NVIDIA NIM container for AI as part of CREATE EXTERNAL MODEL. In this case, we had to use the proxy to change the headers going into the AI model REST endpoint.

Payload

The payload is your "data" you are sending to the REST service when using a method like POST. This is a valid Unicode string in JSON, XML, or text format. There is no size restriction of what SQL Server can send, but the REST service may have a maximum restriction.

The allowed format of the payload depends on what the REST service supports. Some REST services may require a specific JSON document, for example, like AI models, but the documentation for the REST service should provide this.

Response

This is an OUTPUT parameter of the stored procedure. This means when a REST service replies back with data, it does not come in the form of a SQL Server result set. This is why the AI_GENERATE_EMBEDDINGS T-SQL function is so important for AI since it can be integrated with INSERT and UPDATE statements.

For SQL Server, this parameter is a NVARCHAR(MAX) type. The format of the response can be JSON, XML, or text. It is up to the REST service to provide details on what valid responses will look like.

I know you are reading this, and it all sounds like "it depends on the REST service." This is both the beauty and cost of using REST.

Credentials

Some REST services publish their endpoint as public or anonymous. This means there is no authentication required on the endpoint itself or service.

Most do require authentication. And how authentication works depends on the REST service.

For this system procedure, we support a DATABASE SCOPED CREDENTIAL. As you learned for AI models in Chapter 4, that credential could be an API_KEY or Managed Identity.

If you use the option **IDENTITY = 'HTTPEndpointHeaders'**, with the credential, the value, like an API_KEY, will be added to the headers sent to the service.

It is also possible that your REST service is running locally, like a reverse proxy used for Ollama, and has a self-signed certificate. These services are protected with HTTPS but don't require a database scoped credential because any program on the machine is authorized with the local certificate. This was the example I showed you in Chapter 4 with CREATE EXTERNAL MODEL when using Ollama.

Return Values and Errors

The documentation docs a good job to show how the return value of the procedure works and possible errors:

`https://learn.microsoft.com/sql/relational-databases/system-stored-procedures/sp-invoke-external-rest-endpoint-transact-sql#return-value`. says

> *"Execution returns 0 if the HTTPS call was done and the HTTP status code received is a 2xx status code (Success). If the HTTP status code received isn't in the 2xx range, the return value is the HTTP status code received. If the HTTPS call can't be done at all, an exception is thrown."*

The documentation also has some common HTTPS errors you might encounter at `https://learn.microsoft.com/sql/relational-databases/system-stored-procedures/sp-invoke-external-rest-endpoint-transact-sql#retry-count-logic`. This includes the option of a retry count for any transient issues.

To help clarify this, if the REST service returns an error, you won't get a SQL error but will need to investigate the return value of the procedure as an HTTP status code. If the problem relates to SQL Server, you will get a SQL Server error and need to focus on that first *even if the return value is 0.*

For example, if you get this SQL Server error

```
Msg 11558, Level 16, State 202
The @result JSON string could not be parsed. Please check the formatting of
the JSON.
```

this means the REST service is returning an invalid JSON document as the response. In this case the return value could be 0 meaning the service had no HTTP errors but the JSON document fails the policy of sp_invoke_external_rest_endpoint.

Engine Details

There are a few details of how this system procedure works within the engine that are worth understanding:

- The invocation of this procedure is synchronous, so your T-SQL statement must wait until the REST service provides a response.
- SQL worker threads run preemptively when executing this procedure, so any delays will not block others from a SQLOS scheduling point of view.
- Because SQL worker threads run preemptively, we don't want to starve the SQLOS scheduling system, so you cannot have more than 150 worker threads or 10% of maximum worker threads concurrently executing this procedure. If you hit the limit, you will fail with Msg 10928.
- The wait_type you will see when a request is currently executing this procedure is HTTP_EXTERNAL_CONNECTION.

Scenarios to Use REST

Our documentation has a lot of great examples of how to use this procedure with Azure services at `https://learn.microsoft.com/sql/relational-databases/system-stored-procedures/sp-invoke-external-rest-endpoint-transact-sql#examples`. In addition, the community has contributed other examples at `https://github.com/Azure-Samples/azure-sql-db-invoke-external-rest-endpoints`. But these examples were done when this procedure has been used for Azure SQL. Are there any examples for local REST API usage for SQL Server 2025?

This is a great community story. When we announced this capability for SQL Server 2025, some interesting examples started coming out.

Anthony Nocentino from Pure Storage, and MVP, wrote a blog post on how to combine REST API services from Pure Storage with snapshot backups. Learn more at `https://www.nocentino.com/posts/2025-06-27-building-a-snapshot-backup-catalog`.

Anthony Nocentino and Andy Yun from Pure built a very nice demonstration to show how to obtain Pure Storage disk storage performance metric using REST and then combine this with SQL Server metrics to get an end-to-end look at I/O performance. I found this one at `https://github.com/nocentino/accelerate-2025/tree/main/sql-2025-and-pure-storage`.

Andrew Pruski used REST APIs to integrate with the Kubernetes API from within the SQL Server engine. Check out his blog post at `https://dbafromthecold.com/2025/07/31/accessing-the-kubernetes-api-from-sql-server-2025`.

Using REST for AI to Complete the RAG Story

I showed you in Chapter 4 how to use CREATE EXTERNAL MODEL to provide vector searching capabilities with AI models.

But remember the figure I showed you in Chapter 3 that I'll show again as Figure 5-13.

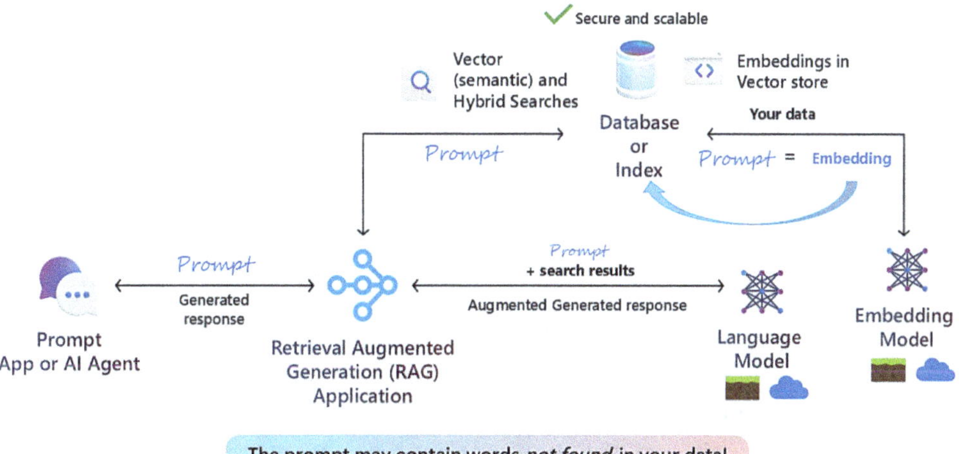

Figure 5-13. *Adding vector search to RAG*

The results from the vector search could be used in any application, even a shopping bot or agent on a website. Let's say you want to take these results and analyze them for gaps or improvements from the perspective of someone *working inside* the AdventureWorks company. To do that we need to send these results to a chat completion model. We could do this in our application, or we could do this **inside SQL Server**.

Let's take a look at how to use **sp_invoke_external_rest_endpoint** to take the results of the vector search I showed you in Chapter 4 in the section titled "Getting Started with Vectors" and augment the prompt with a *chat completion model.*

Prerequisites for the Example

- All the scripts in these examples can be found in the **ch5 - Developer\ sp_invoke_external_rest_endpoint** folder.
- You must have completed the example in Chapter 4 in the section titled "Getting Started with Vectors" to perform a vector search against the AdventureWorks database using Azure AI Foundry.

Note You can use whatever chat completion model you like that can be accessed by a REST endpoint, ground or cloud, or uses local ONNX model support. Your results may vary from this example depending on what model you choose. Having the model of your choice is one of the powerful capabilities of SQL Server 2025.

This example would have required you to restore the AdventureWorks sample database, enable the **external rest endpoint enabled** server configuration option, and run all the scripts to execute a vector search.

- You would also already have created a database scoped credential based on access to Azure AI Foundry.
- To reuse that credential, create a deployment of a chat completion model (I used **gpt-4o**) in the Azure OpenAI services that you used for the embedding model you used in Chapter 4. This allows you to use the same API key. Use this documentation page as guidance: `https://learn.microsoft.com/azure/ai-foundry/openai/how-to/create-resource`.

Note You will see in Chapter 6 of the book you can alternatively use a Managed Identity for authentication.

Example to Extend RAG with Vector Search

For this example there is only one script to use called **chatcompletions.sql**. This script is one batch that uses many variables. You could have deployed this in a stored procedure to abstract the code. All of this is executed in the context of the AdventureWorks database.

I've broken down the script into steps, so let's evaluate each step and *then execute the entire script* to see the results:

1. **Store the results of the vector search in a temp table.**

```
-- STEP 1: Store the results of vector search in a temp table
DROP TABLE IF EXISTS #ProcResult;
CREATE TABLE #ProcResult
(
    ProductName         NVARCHAR(200)  NULL,
    ProductDescription  NVARCHAR(MAX)  NULL,
    StockLevel          INT            NULL
);
INSERT INTO #ProcResult (ProductName, ProductDescription,
StockLevel)
EXEC find_relevant_products_vector_search
@prompt = N'Show me stuff for extreme outdoor sports',
@stock = 100,
@top = 20;
```

   We need to format this as JSON, so store the results first in a temp table.

2. **Our chat completion model expects JSON, so let's format our result set into JSON.**

```
-- STEP 2: Convert the result set to JSON
DECLARE @resultSetJson NVARCHAR(MAX);
```

```
SELECT @resultSetJson =
(
    SELECT ProductName, ProductDescription
    FROM #ProcResult
    FOR JSON PATH, INCLUDE_NULL_VALUES
);
```

This is where it is so nice to have JSON functions that easily convert results into a JSON document.

3. **Create parameters for chat completions and REST API.**
   This is where it gets a bit interesting.

```
-- STEP 3: Build Chat Completions parameters
DECLARE @escapedJson   NVARCHAR(MAX) = STRING_ESCAPE(@
resultSetJson, 'json');
DECLARE @url NVARCHAR(MAX) = N'https://<azureai>/openai/
deployments/gpt-4o/chat/completions?api-version=2025-01-01-
preview';
DECLARE @credential NVARCHAR(4000)  = N'https://<azureai>.openai.
azure.com';  -- DB scoped credential name
DECLARE @userPrompt NVARCHAR(MAX)   = N'Are these good products for
someone who is an extreme sports enthusiast. Anything missing?';
DECLARE @payload NVARCHAR(MAX);
SET @payload =
N'{
  "messages": [
    { "role": "system", "content": "You are a helpful assistant
    that analyzes small SQL result sets about products from
    an outdoor sports company. Be concise, structured, and
    actionable." },
    { "role": "user",   "content": "'+@userPrompt +'"},
    { "role": "user",   "content": "'+@escapedJson+ '"}
  ],
  "temperature": 0.7
}';
```

@escapedJson is to ensure we have valid JSON text.
@url is the location endpoint for your chat completion model.
Figure 5-14 shows an example of how to get the URL and API_KEY for gpt-4o in Azure AI Foundry.

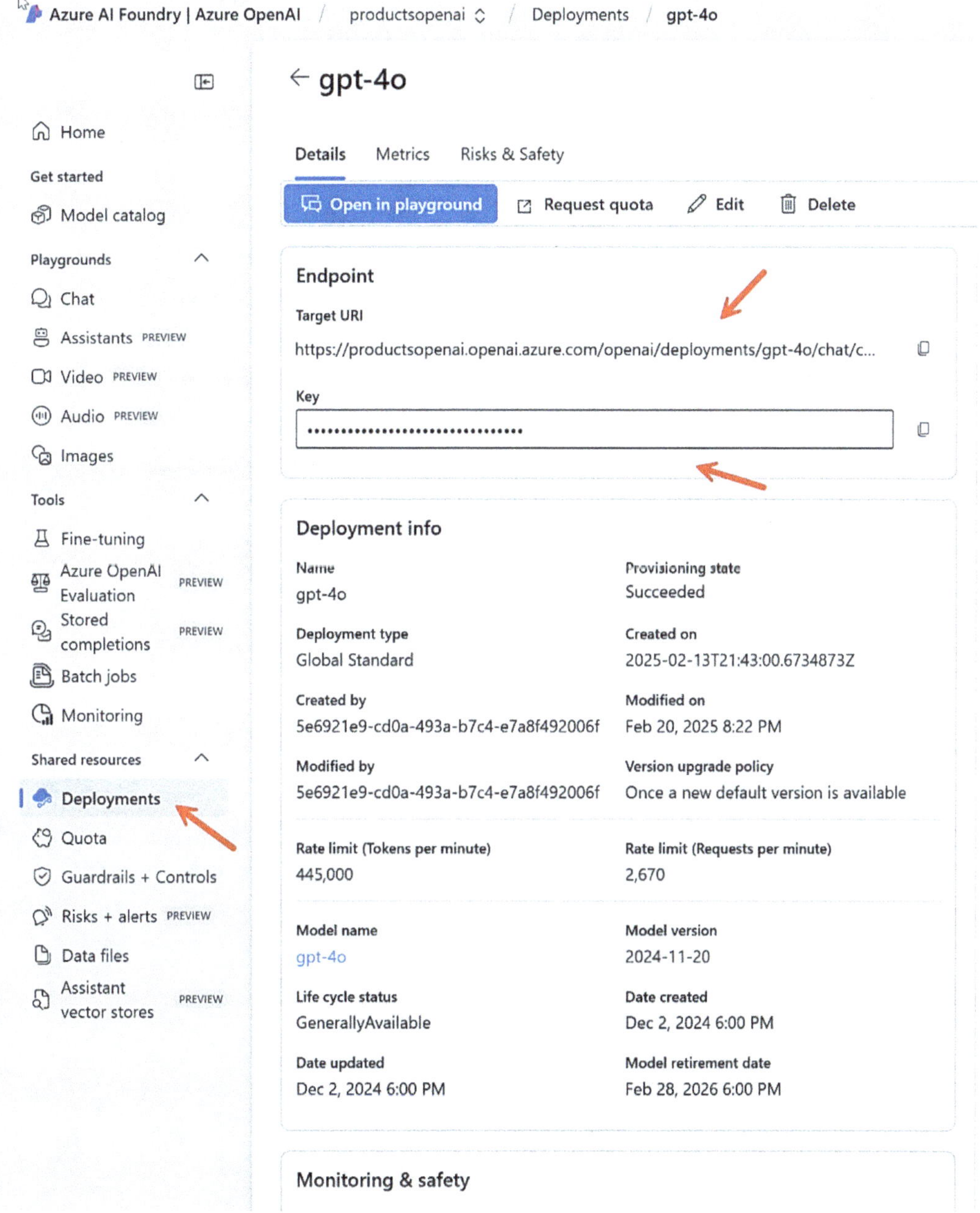

***Figure 5-14.** Obtaining the endpoint for an AI model in Azure AI Foundry*

@credential is the database scoped credential you created for the embedding model in Chapter 4. Again, if you use a different REST approach, this credential may be different, but you will need a separate step before this to create it.

@userPrompt is the "user" message to send to the model, but we will send two user messages for this scenario: one for this "instruction" different than the system prompt and one for the result set.

@payload is a combination of text, prompts, user prompts, and parameters that the AI model accepts. Notice we use a "system prompt" as instructions on how the model should "behave." This affects the response of the model. The payload contains two user prompts for further instructions and the result set in JSON. Other parameters can be used to control model behavior as I described to you in Chapter 3. In this case we will use a **temperature** of 0.7 (gpt-4o accepts 0–2.0 with 1.0 being the default). Remember the higher the number, the more creative the model responds, but this can also result in inconsistent and even unpredictable results.

4. Send a message to the chat completion model.
   After all that work here is the code to send a REST API message to the model:

```
-- STEP 4: Execute the chat completion
DECLARE @statusCode INT;
DECLARE @response NVARCHAR(MAX);
EXEC @statusCode = sp_invoke_external_rest_endpoint
    @url        = @url,
    @method     = N'POST',
    @payload    = @payload,
    @credential = @credential,
    @response = @response OUTPUT,
    @timeout    = 120;
```

Depending on your model location and capabilities, this can come back very quickly or even take seconds or even minutes to return. You might consider adding error handling here if @statusCode is > 0. One I've seen when using a cloud service like Azure AI Foundry is HTTP status code 429, which means "Requests to the OpenAI API have been throttled. Please retry after some time." You will need to increase resources or delay sending another request.

5. **Extract out the response message.**

```
-- STEP 5: Extract out the response message
SELECT c.content
FROM OPENJSON(@response, '$.result.choices')
WITH ( content NVARCHAR(MAX) '$.message.content' ) AS c;
GO
```

The @response is a JSON-formatted message for which the "content" has your actual model response message. But the overall response has other interesting information you might want to look at including errors.

For example, if you get HTTP status code 400, your payload may be incorrect. The @response would have the detailed error of what was incorrect in it.

In this case, we got back something very interesting like the following text:

Analysis of Products for Extreme Sports Enthusiasts: The products listed seem well-suited for extreme sports enthusiasts, particularly in mountain and road biking. Here's a breakdown: --- #### **Strengths:** 1. **Mountain Bikes:** - Products like the *Mountain-100 Silver*, *Mountain-200 Silver*, and *Mountain-300 Black* are robust and designed for challenging terrain and competition. - Different sizing options (38, 40, 42) cater to a variety of riders. 2. **Road Bikes:** - The *Road-450 Red* series offers aerodynamic, multi-sport capabilities and hill-conquering gearing, which is excellent for extreme road cycling. 3. **Components:** - *HL Mountain Wheels* (front

and rear) provide high-performance replacements for extreme mountain biking. - *Handlebars* (LL and ML) are versatile, with ML specifically built for downhill challenges. --- #### **Gaps / Potential Missing Items:** 1. **Protective Gear:** - No helmets, gloves, or body armor are listed—essential for extreme sports enthusiasts to ensure safety. 2. **Specialized Accessories:** - Items like bike lights, GPS trackers, hydration systems, or bike racks are missing but are critical for extreme conditions and long rides. 3. **Footwear and Apparel:** - No mention of specialized biking shoes, jerseys, or weather-resistant clothing. 4. **Other Extreme Sports Gear:** - The list focuses solely on biking. For broader appeal to extreme sports enthusiasts, gear for climbing, kayaking, or snow sports could be considered. 5. **Maintenance Tools:** - No tools or repair kits for on-the-go fixes in extreme conditions. --- ### Recommendations: - **Expand Product Categories:** Include safety gear, apparel, and accessories for a comprehensive lineup. - **Diversify Sports:** Broaden offerings to cater to other extreme sports like climbing, skiing, or water sports. - **Promote Bundles:** Offer bike + safety gear or bike + accessory bundles tailored for extreme sports users. These additions could significantly enhance the appeal for extreme sports enthusiasts.

Notice how the model provides all types of advice on the results and how they match up to provide the right choices for extreme sports enthusiasts. Notice that this model knows the term "extreme" similar to our use of embeddings for vector searching.

The possibilities are endless to use this capability. And for this scenario, like vector searching, we performed all this with T-SQL **and inside the security boundary of the SQL Server engine**.

The Modern SQL Developer

SQL developers are back! Bring in semi-structured documents with the new JSON support. Use the power of RegEx with T-SQL instead of having to put this logic in your application or build a SQLCLR procedure. Build new event-driven architectures using

Change Event Streaming. And then go further by connecting to REST services, including AI chat completion models, ground or cloud. Combine this with the new AI vector search, and you have a powerful platform to build modern data applications.

Stay in touch with the latest resources for SQL developers at `https://aka.ms/sqldev` or keep up with the latest news at the Azure SQL Devs' Corner at `https://devblogs.microsoft.com/azure-sql`.

Keep your SQL Server 2025 journey going by learning in the next chapter how to connect your SQL Server to Azure Arc.

CHAPTER 6

Connecting to the Cloud with Azure Arc

SQL Server runs on any platform you need, ground to cloud to fabric. Regardless of where you run SQL Server, a common solution for our customers is to *connect* SQL Server to the cloud, wherever SQL Server is running. The primary method to connect SQL Server to the cloud is to connect your deployments to Azure through **Azure Arc**. Azure Arc–connected SQL Server deployments are the Microsoft definition for a *hybrid* SQL Server.

In this chapter, I'll explain the *what, how, and why* about Azure Arc and the value of using Arc to connect SQL Server to Azure. Before you read this chapter further, you should know that there is no requirement to use Azure Arc for most new features of SQL Server 2025. The only features that require Azure Arc are Fabric Mirroring, which you will learn more about in Chapter 8 of the book, and Microsoft Entra, which is covered in this chapter. Having said that, Azure Arc can provide "value-added" services and capabilities. You will learn all about these in this chapter.

In this chapter I have provided examples for you to go through to learn how to use Azure Arc with SQL Server. For these examples, you will need an Azure subscription and the ability to connect your SQL Server to Azure, either directly connected to the internet or through a proxy. Your company or organization may already have an Azure account for you to use or a subscription. However, if you need your own, start at `https://azure.microsoft.com/get-started`. You can find all the scripts used in this chapter in the **ch6 – Azure Arc** folder.

B. Ward, *SQL Server 2025 Unveiled*, https://doi.org/10.1007/979-8-8688-1847-9_6

What Is Azure Arc?

From this point forward in the chapter, I may use the term **Arc** to mean **Azure Arc**. Why the term *Arc*? You might be thinking Arc in terms of a curved line connecting two objects. But Arc also means "a discharge of electricity across a gap …", and that is Arc is all about – connecting resources anywhere to the Azure cloud that are *not running in Azure*.

Azure Arc is the name of a family of hybrid capabilities including servers, Kubernetes clusters, data services, and SQL Server. One great resource to get started on everything with Azure Arc is **Jumpstart**, which you can find at `https://aka.ms/azurearcjumpstart`.

Azure Arc requires two key overall components:

- Resource providers in Azure to manage Azure resources. I'll list later in the chapter which ones you need to use Arc with SQL Server.
- Software agents that run in virtual machines, Kubernetes clusters, or servers.

From the perspective of SQL Server, the most important component to enable Arc is the **Azure Arc Connected Machine Agent**.

Figure 6-1 shows the architecture of the Azure Connected Machine Agent (this diagram comes from our documentation at `https://learn.microsoft.com/azure/azure-arc/servers/agent-overview`).

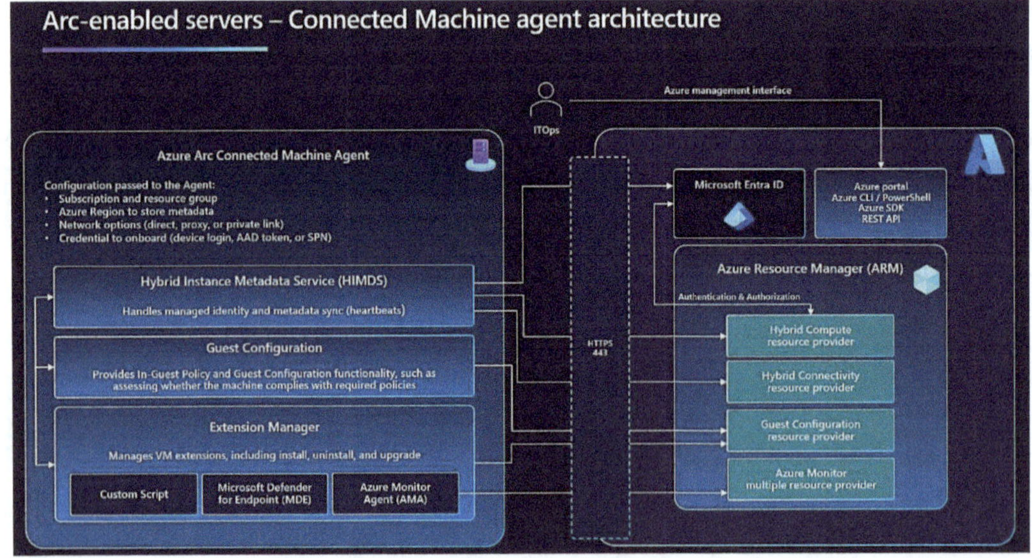

Figure 6-1. *The Azure Arc Connected Machine Agent*

The agent itself is a software program that runs as a Windows service or Linux daemon program. One of its main purposes is to register and maintain the "machine" as an Azure resource. You see this in the Azure Portal as an **Azure Arc Machine** like in Figure 6-2.

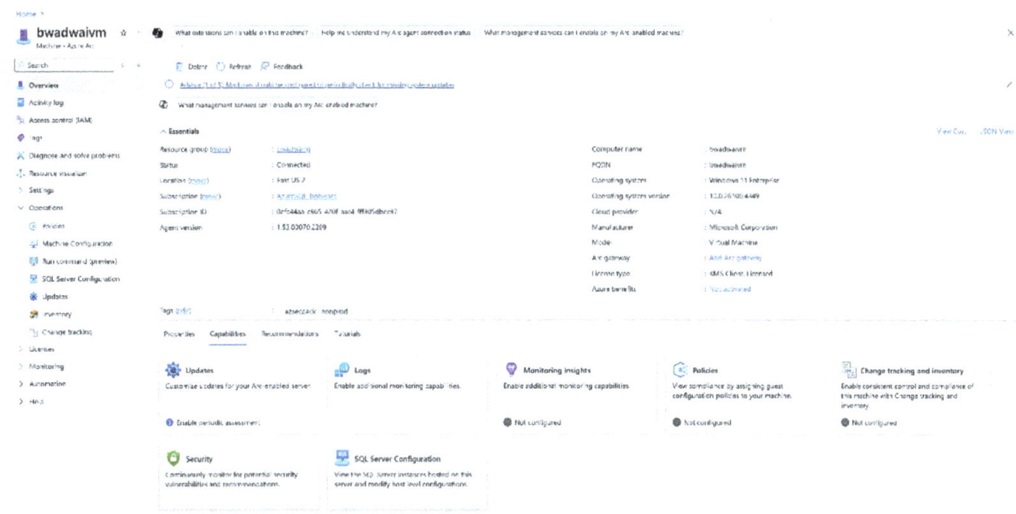

Figure 6-2. *Azure Arc Machine resource*

This is similar to when you deploy SQL Server on an Azure Virtual Machines; the VM itself appears as a separate resource. But remember your "machine" with Arc is *not running* in Azure. We simply track metadata about your machine in Azure as an Azure resource. Some users will deploy Azure Arc without SQL Server to gain hybrid capabilities for the *machine* including but not limited to

- Free inventory of the server
- Azure Update Manager to manage and automate OS updates
- Monitoring operating system performance

The other major capability for the Azure Connected Machine Agent is to support *extensions*. Azure Virtual Machines has the same concept. One of the most common Arc extensions is the Azure Monitor Agent (AMA), which can be used to collect rich performance and event data and analyzed in a Log Analytics Workspace in Azure. For SQL Server, the agent also supports the **Azure extension for SQL Server**. This is a key software component to support a SQL Server enabled by Azure Arc.

SQL Server Enabled by Azure Arc

If the goal of Azure Arc is to connect resources to Azure, then it makes sense that the goal of SQL Server enabled by Azure Arc is to connect a SQL Server *not running in Azure* to the Azure cloud. That SQL Server could be running in your private cloud (on-premises) or in other public clouds. The goal is to provide value-added services using the power of Azure to your SQL Server instance.

The agent software that enables the value-added services specific to SQL Server is the **Azure extension for SQL Server**.

What Is the Azure Extension for SQL Server?

Here is the architecture for the extension for SQL Server in Figure 6-3 (this diagram comes from our documentation at `https://learn.microsoft.com/sql/sql-server/azure-arc/overview#architecture`).

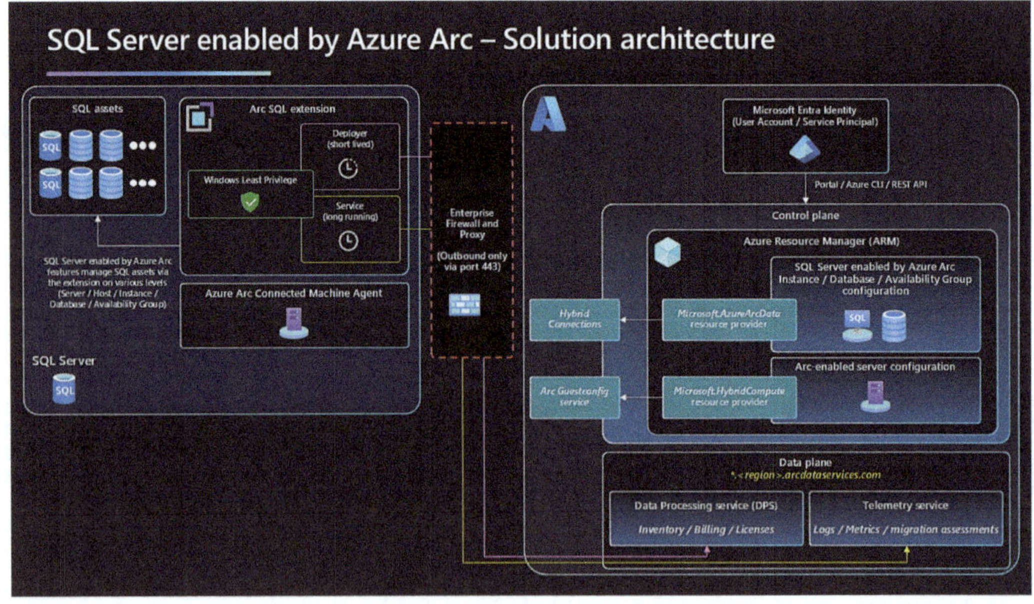

Figure 6-3. *SQL Server enabled by Azure Arc architecture*

As you can see in the architecture diagram, the extension uses outbound port 443 to communicate with a *data plane* in Azure to support some of the value-added services such as inventory, billing, licensing, metrics, assessments, and more. Let's look at more details of these features and then see how to get started with examples. You also have the ability to run the agent for the extension with a least privileged account.

Enabling Hybrid Scenarios with Azure Arc

Here is a brief look at all the features that SQL Server enabled by Azure Arc provides for a hybrid SQL Server.

Note We have the ability to update Azure Arc frequently, so keep track of any new updates with our release notes at `https://learn.microsoft.com/sql/sql-server/azure-arc/release-notes`. The features you see below can vary by license type and operating system. And in some cases, we need features in a new version of SQL Server to enable a feature. Check our documentation at `https://learn.microsoft.com/sql/sql-server/azure-arc/overview#feature-availability-depending-on-license-type`. Scroll down after this list to see these differences.

As you read about these features, keep in mind that one of the biggest values for Azure Arc is viewing and managing your SQL Server instances *at scale,* regardless of whether they are deployed on-premises or in other clouds. In addition, some of these features are free by just connecting to Azure. Others require some cost based on Azure resource usage.

Inventory

Azure Arc provides a centralized inventory of all connected SQL Server instances. This includes detailed metadata such as instance name, version, edition, number of cores, and host operating system. This visibility helps streamline asset management and compliance tracking across hybrid environments.

You can view a complete list of databases across all Arc-enabled SQL Server instances. This feature supports cross-instance queries, enabling insights into backup status, encryption, and other key attributes, which is essential for maintaining data integrity and operational oversight. I picked this as the first feature because it is *free*.

In addition, you can query your inventory with Azure Resource Graph Explorer; you can query across all Arc-enabled SQL Server instances. This allows you to answer questions like "How many SQL Server 2014 instances do I have?" or "Which instances are running on Linux?" and visualize the results in dashboards.

Licensing

SQL Server is typically licensed by a single paid license or through a contract like Software Assurance. Azure Arc enables a new license type called pay-as-you-go (PAYG). PAYG allows you to pay for a SQL Server license using your Azure subscription based on the usage of SQL Server. You will see an example of this later in the chapter.

Extended Support Updates (ESU)

Extended Support Updates allows you to obtain support for older versions of SQL Server that are out of mainstream support. ESU requires a contract with Microsoft and usually an up-front cost. Using Azure Arc, you can use a PAYG model to pay for the ESU license and receive ESU automatically to SQL Server.

Best Practices Assessment

This feature evaluates your SQL Server configurations against Microsoft's recommended best practices. It provides actionable insights to improve performance, security, and compliance, helping you proactively address potential issues before they impact operations.

Microsoft Entra ID Authentication

Formerly known as Azure Active Directory, Microsoft Entra ID enables modern, centralized identity and access management for SQL Server. This feature enhances security by eliminating the need for local credentials and supports Managed Identity authentication for Azure services. There are new capabilities for this feature in SQL Server 2025, which you will learn about later in this chapter.

Azure Policy Integration

Azure Policy allows you to enforce governance rules across your SQL Server estate. You can define and audit compliance with policies related to configuration, security, and operational standards, ensuring consistent management across environments.

Azure Monitor Integration

Azure Monitor provides deep observability into SQL Server performance and health. It collects telemetry data, supports alerting, and integrates with dashboards to help you detect and respond to issues quickly.

Azure Log Analytics Integration

Log Analytics enables advanced querying and analysis of logs from Arc-enabled SQL Servers. This helps in identifying trends, troubleshooting issues, and maintaining operational visibility across your SQL infrastructure.

Microsoft Defender for SQL

This security feature offers advanced threat protection for your SQL Server instances. It detects and alerts on suspicious activities, helping you safeguard sensitive data and meet compliance requirements.

Automated Local Backups

You can configure automated local backups for Arc-enabled SQL Server instances. This ensures regular data protection without manual intervention, supporting business continuity. This also includes the ability to perform point-in-time resources from these backups.

Automated Patching

Azure Arc supports automated patching of SQL Server instances, reducing administrative overhead and ensuring systems remain secure and up to date. Today this feature only supports security updates.

Monitoring

Azure Arc provides basic performance monitoring including client connection details viewable in the Azure Portal.

Always On Management

You can view and manage Always On SQL Server Failover Cluster Instances (FCIs) directly from the Azure Portal. This includes monitoring node status, network configuration, and associated databases, simplifying high availability management. Azure Arc enables visibility and control over Always On Availability Groups. You can monitor health, view configurations, and initiate failovers, all from a centralized interface.

Now that you have an idea of what is possible with Azure Arc, let me show you some examples of these features in action.

Getting Started with Azure Arc

Let's see how to get started using Azure Arc to set up the following scenarios:

- View and manage SQL Server in the Azure Portal including inventory. This includes gaining custom insights from Azure resource graphs.
- Show how to change the license type for SQL Server.
- Show how to configure ESU licensing.
- Log in to SQL Server using Microsoft Entra.
- Secure access to Azure OpenAI using Microsoft Entra and a Managed Identity.

I will show you how to set up Azure Arc using the built-in experience of SQL Server 2025 setup, which you can read more about at `https://learn.microsoft.com/sql/database-engine/install-windows/install-sql-server-from-the-installation-wizard-setup`.

You can see other options to set up Azure Arc at `https://learn.microsoft.com/sql/sql-server/azure-arc/deployment-options`. For Linux, you will use the az CLI to install Azure Arc. VMWare has some special options, which you can read at `https://learn.microsoft.com/sql/sql-server/azure-arc/overview#support-on-vmware`.

Example: Deploying Azure Arc Through SQL Server 2025 Setup

Let's see the details of how to deploy Azure Arc and the scenarios I listed above.

Steps to Deploy and Prerequisites

Here are the following steps and perquisites to deploy Azure Arc through SQL Server 2025 setup.

Note Arc doesn't work on Azure VM (this is because there already is the IaaS Agent). But you can make the VM *think* it is not registered in Azure by using the docs at `https://learn.microsoft.com/azure/azure-arc/servers/plan-evaluate-on-azure-virtual-machine`. That is what I did for this example.

---

1. You will need the following (see a complete detailed explanation at `https://learn.microsoft.com/sql/sql-server/azure-arc/prerequisites`) before you run SQL Server 2025 setup:

    a. An Azure subscription.

    b. A resource group deployed. Note the region for the resource group. Although not required I usually deploy Azure Arc in the same region as the resource group.

    c. A Microsoft Entra login with specific permissions including Read permission in the subscription but also with specific permissions in the resource group. You can be a Contributor or Owner to make it simpler. Specific permissions are called out in Prerequisites documentation listed above.

    d. These Azure resource providers must be registered: **Microsoft.AzureArcData** and **Microsoft.HybridCompute.** Learn more at `https://learn.microsoft.com/azure/azure-resource-manager/management/resource-providers-and-types#register-resource-provider`.

2. Run SQL Server 2025 setup. I used Enterprise Developer Edition for this exercise.

3. One of the first screens you get during setup looks like Figure 6-4.

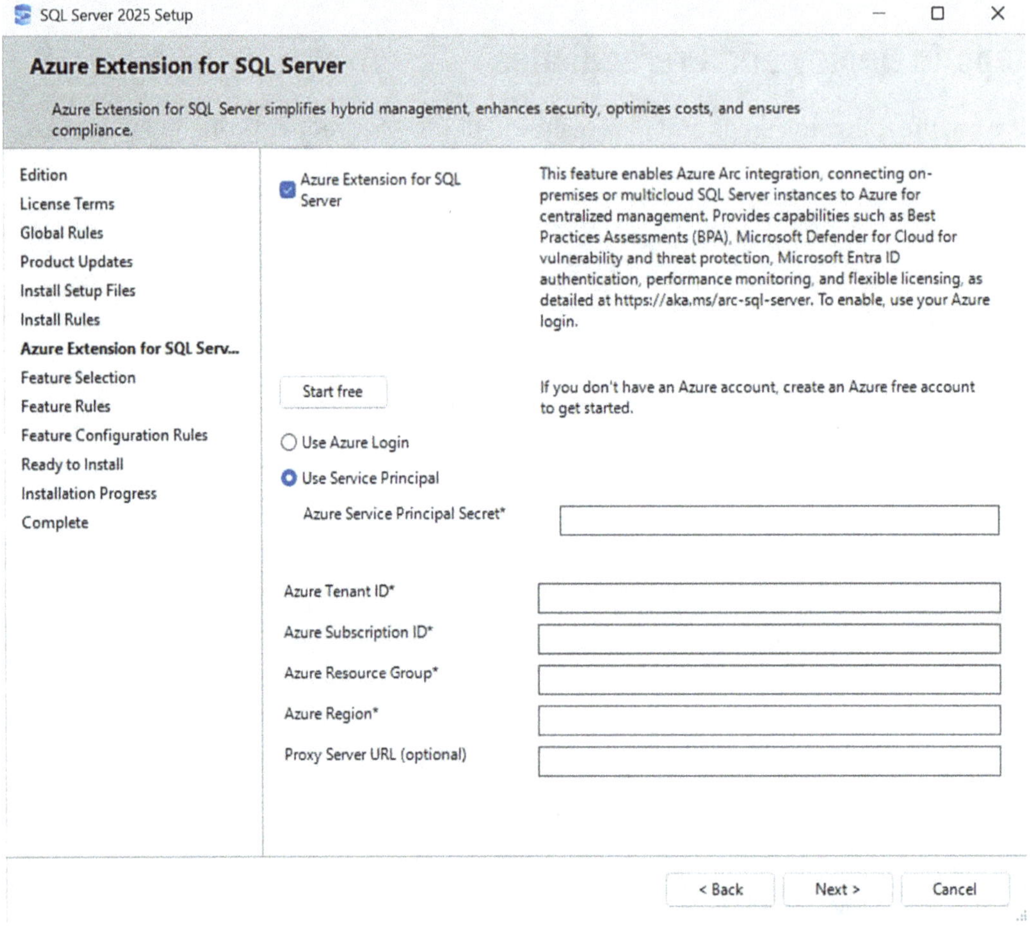

Figure 6-4. *The Azure extension for SQL Server setup screen*

In this case, I have not deployed an Azure Arc agent, so these steps will both deploy the Azure Connected Machine Agent and the SQL Server extension.

Let's break down your inputs for this screen to start the process. (If you didn't know, you can uncheck the option for Azure extension for SQL on this screen to skip the setup. The * marks fields that are required.)

Use Azure Login

The default is Use Service Principal, which you can select as an option. I prefer Azure login, so when I click this option, it brings up a dialog box for me to choose my Azure login. I choose my login from Microsoft, or you may need to put in your Azure account based on your organization's email address or for your own Azure account the corresponding email address.

Azure Tenant ID

This is the Microsoft Entra Tenant for the service principal (which you will need to provide). If you use Azure login, it is automatically filled in and is the Microsoft Tenant ID for the overall tenant.

Azure Subscription ID

Many Azure users are members of multiple subscriptions. Here you can use a drop-down to choose which Azure subscription you want to use to deploy the Azure Arc resource.

Azure Resource Group

This is the name of the Azure resource group where the Azure Arc resource will be deployed. You cannot create a new one here, so this must use an existing resource group. Only resource groups for the Azure subscription you selected are shown.

Azure Region

This is the Azure region in which the Azure Arc resource will be deployed. It is typically the same region as the resource group you are using but does not have to be.

Proxy Server URL

If you have to use a proxy server to connect inbound and outbound from the server where SQL Server is being deployed, you can specify the URL to use in this field.

The setup of the agents does not happen until you have gone through the steps in the process to install.

Exploring Azure Arc Resources

Let's see what has been installed to know your SQL Server is now connected with Arc.

Note Once SQL Server setup is successful, it may take several minutes for all Azure Arc resources to be visible in the Azure Portal.

First, if you look at installed programs on Windows, you will see two new programs as seen in Figure 6-5.

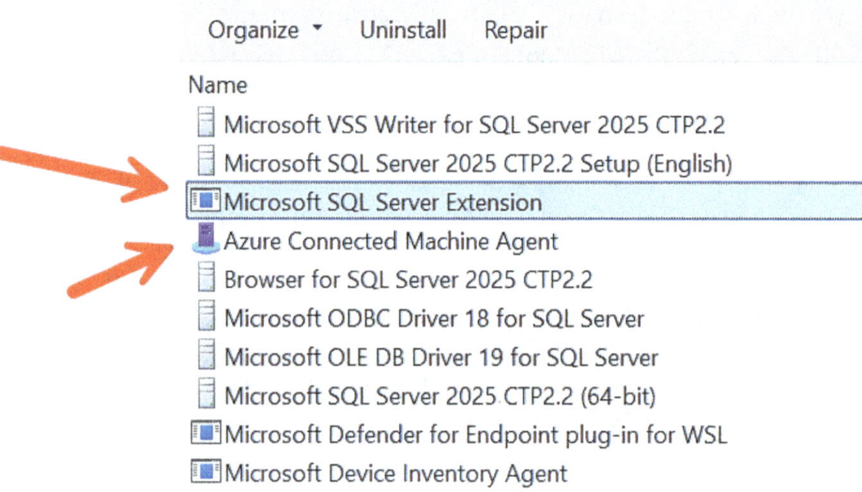

Figure 6-5. *Programs installed on Windows for Azure Arc*

These programs are the agents mentioned earlier in this chapter running as Windows services but also showing up as installed Windows applications. The Microsoft SQL Server Extension is another name for the Azure extension for SQL Server. You can also see these programs installed by looking at Windows Services (using the **services.msc** application) called the **Windows Azure Guest Agent** and **Microsoft SQL Server Extension Service**.

In the Azure Portal you can look at the resource group you used during SQL Server 2025 setup to see the two resources like in Figure 6-6.

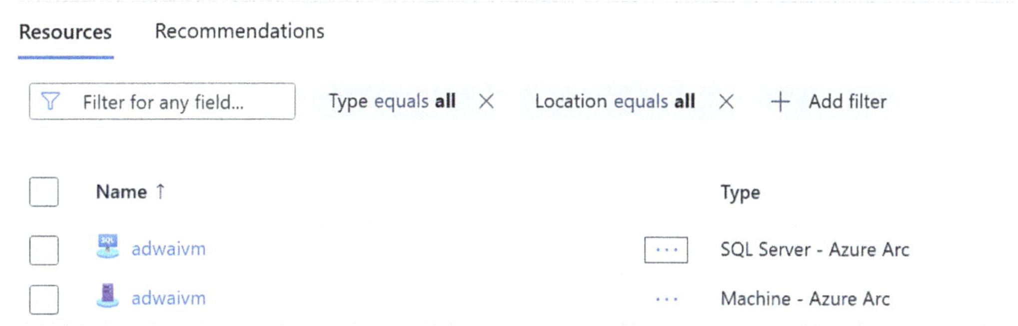

Figure 6-6. *Azure Arc resources after SQL Server 2025 setup*

The **Machine – Azure Arc** resource is for resources associated with the VM (machine) and Windows. The **SQL Server – Azure Arc** resource is associated with all SQL Server Arc capabilities. You can have a Machine – Azure Arc resource on its own, but the SQL Server – Azure Arc resource must have an Azure Arc machine resource present.

If you click the SQL Server – Azure Arc resource, you will see an overview of the resource like in Figure 6-7.

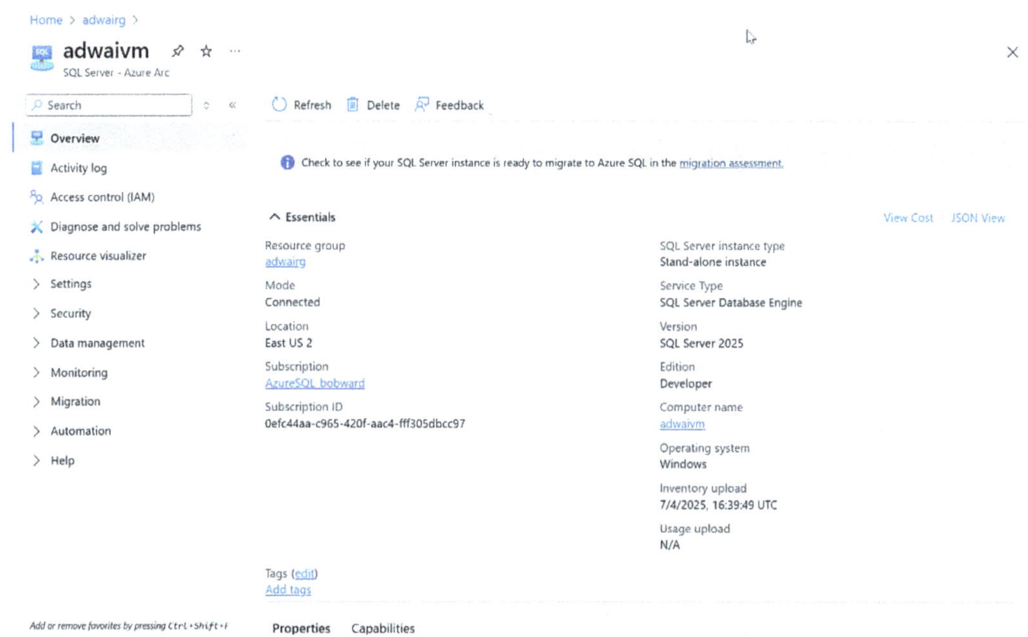

Figure 6-7. *The SQL Server – Azure Arc overview*

Here you will find basic information about the Azure resource (remember the VM and SQL Server are not running in Azure, but there is a resource stored in Azure to keep metadata about the machine and SQL Server) and the SQL Server instance deployment including Version, Edition, Computer name, Operating system, and datetime information on uploads for inventory and usage.

Scrolling down on this screen, you see other information related to SQL Server and the status of the Azure extension.

As I highlighted earlier in this chapter, Azure Arc provides several value-added services to your SQL Server deployment. You can find access to many of these in the left-hand menu when viewing the SQL Server - Azure Arc resource. In the next section, I'll show you how to secure your SQL Server with Microsoft Entra. But let me first show you other capabilities I think you will find valuable.

First, one of the simple yet elegant advantages of using Azure Arc is *free inventory* of your installation of SQL Server. You saw in the overview page in Figure 6-7 a small example of this. Since this is a SQL Server, you can also see a list of your databases and details about them. Figure 6-8 shows a list of databases on my SQL Server instance.

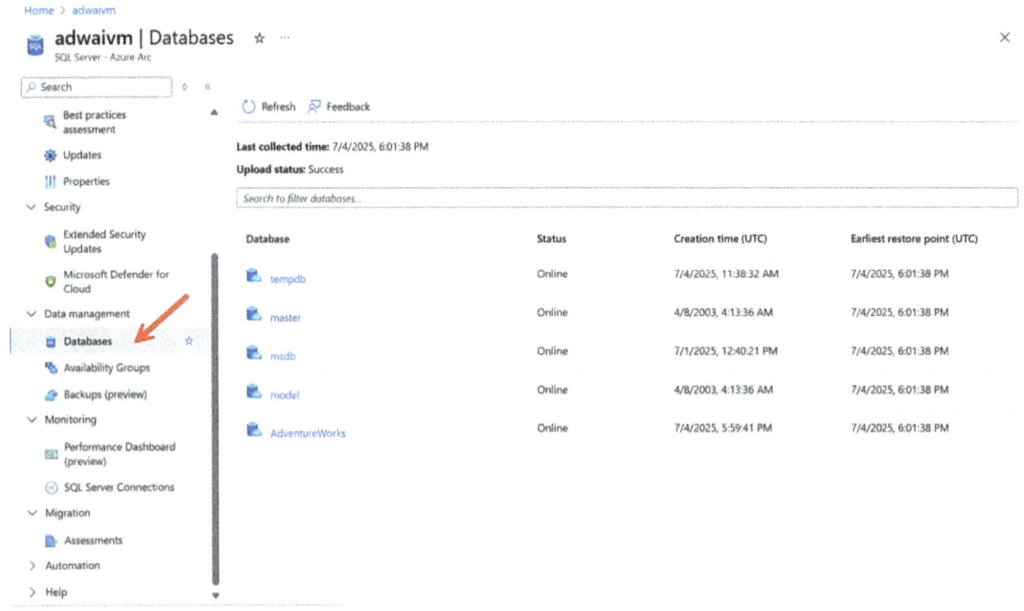

***Figure 6-8.** Inventory of databases for SQL Server enabled by Azure Arc*

If you click one of the databases, in this case the AdventureWorks database I have deployed, you can get details of the database like in Figure 6-9.

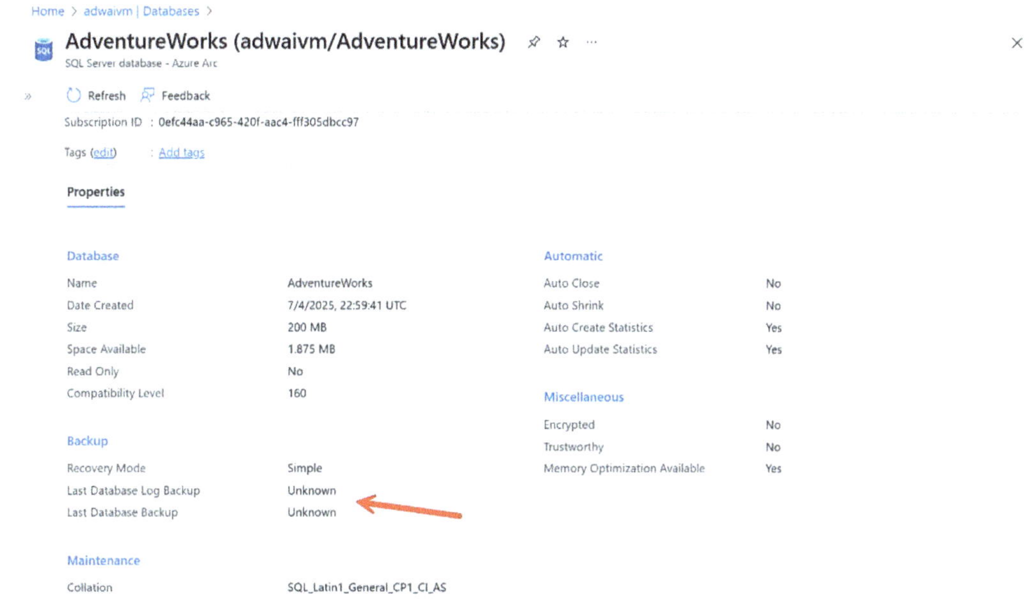

Figure 6-9. *SQL Server Azure Arc database details*

There are some great details to see on this page about the database. You can see highlighted this database has never been backed up.

While looking at the properties of an individual database is interesting, the magic of using Azure Arc is *inventory at scale*. Behind the scenes you can use Azure Resource Graph to run queries to view the inventory of properties of the instance and databases across all SQL Server deployments connected with Azure Arc. And the best part? It is free! In Figure 6-10, I used Azure Copilot to help generate a resource graph query to find a histogram of all dbcompat levels across all SQL Server deployments.

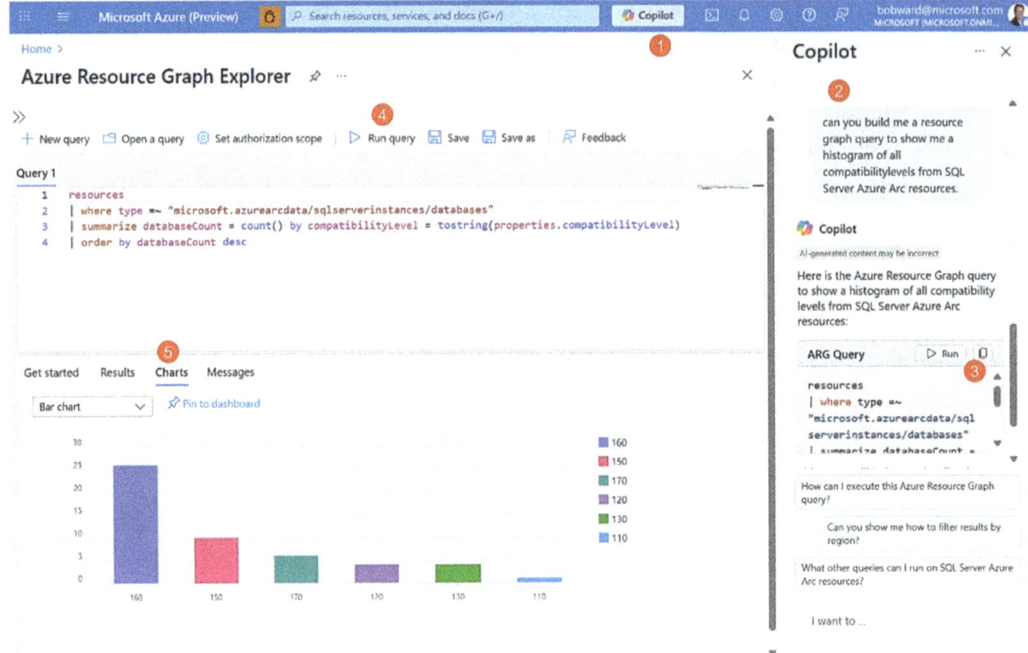

Figure 6-10. *Looking at dbcompat across all SQL Servers using Copilot*

Let's look at each step on how I did this:

1. I selected Azure Copilot at the top of the page, which brings up a *sidecar chat.*
2. I put in a prompt like you see in the figure.
3. Copilot came back with a suggested Azure Resource Graph query. I selected to Run.
4. I was presented with the query and hit Run.
5. I got the results and selected Charts to see the histogram.

License Types

Another interesting reason to use Arc is for *pay-as-you-go* (PAYG) licensing. PAYG allows you to license SQL Server per core per hour. If you shut down SQL Server, you don't incur any licensing costs. This can be a very compelling choice for infrequently used

SQL Server instances and similar to using SQL Server on Azure Virtual Machines. On the overview page you can choose License Type and see a page like Figure 6-11 to choose among various licensing types.

License Type *

Specify the SQL Server edition and license type you are using on this machine. Learn more

- **Pay-as-you-go**
  I want to license my production environment on this server with Enterprise or Standard edition using pay-as-you-go ("PAYG")
- **License with Software Assurance**
  I have a production environment on this server with Enterprise or Standard edition covered by Software Assurance or SQL subscription ("Paid")
- **License only**
  I use other license types on this server with Evaluation, Developer, Express edition or a SQL license without Software Assurance ("LicenseOnly")

Figure 6-11. *Choosing a license type with Azure Arc*

Finally, if you are using an older version of SQL Server that is eligible for Extended Support Updates (ESU), you can use Arc to pay for ESU. Figure 6-12 shows you where you can configure for your Arc deployment for ESU.

Extended Security Updates

To activate ESU subscription, your eligible SQL Server version on this machine must be covered by Software assurance or use Pay-as-you-go subscription. Learn more

Subscribe to Extended Security Updates
By selecting this option, you understand and confirm that
- To qualify for ESU subscription, you have purchased ALL prior years of ESU coverage.
- If you have more than one eligible out-of-support SQL Server version installed on this machine, each will be billed separately based on the installed edition.
- If this machine has a standalone installation of SQL Server Analysis Services (SSAS), SQL Server Integration Services (SSIS), SQL Server Reporting Service (SSRS) or Power BI Report Server (PBIRS), it will be billed for ESU based on the installed edition and version.
- Based on your current configuration and the day of activation, an estimated one-time bill-back charge of **$0** will be applied immediately, estimated hourly charge of **$0** will follow. These costs will appear as separate line items in invoicing.
- If you are using a Azure dev/test subscription or if your SQL instances qualify for the HADR benefit, the charges will be nullified.

Unsubscribe from Extended Security Updates
By selecting this option, you confirm that you wish to terminate Extended Security Updates for the SQL Server 2012 instances on this machine, and immediately stop the ongoing charges.

Physical core ESU license

Specify if this virtual machine is using a SQL Server physical core ESU license that exists and is activated for a tenant, subscription, or resource group to which this virtual machine belongs. If these conditions are not met, the virtual machine will use the specified license type. Learn about SQL Server License management.

Apply a physical core ESU license, if one exists, to this virtual machine

There are currently no physical core ESU licenses that would apply to this host.

Figure 6-12. *ESU with Azure Arc*

Get more information at `https://learn.microsoft.com/sql/sql-server/azure-arc/extended-security-updates`.

Now that you have seen some fundamental reasons to use Azure Arc with SQL Server, let's look at a scenario to secure your SQL Server with Azure Arc and Microsoft Entra.

Securing SQL Server with Microsoft Entra

SQL Server has a proven security system including authentication, encryption, data protection, and auditing. However, Microsoft Entra (formerly Azure Active Directory) offers compelling capabilities that can enhance the SQL Server security experience. In SQL Server 2022, we introduced the ability to use Microsoft Entra for SQL Server, even when it is not running in Azure. Azure Arc empowers this ability.

Microsoft Entra is a managed service for authentication for all types of applications and services. Think of this as a Microsoft managed set of domain controllers that you can use to create your own directory, users, groups, and authentication schemes. Unlike using Windows Active Directory, Microsoft Entra provides a central, consistent security platform for SQL, ground to cloud to fabric. In addition, Entra provides unique capabilities including login with Multi-factor Authentication (MFA) and *passwordless* security with managed identities.

Using Azure Arc, Microsoft Entra is tightly integrated into the SQL Server security ecosystem. Microsoft Entra users and objects such as a Managed Identity are now *securables* for SQL Server authentication and permissions. The SQL Server engine has all the metadata it needs in the Windows registry or mssql.conf file (created and maintained by the Azure extension for SQL Server) to know how to connect to Microsoft Entra to authenticate any Microsoft Entra account that has been configured to use SQL Server.

Example: Connecting to SQL Server with Microsoft Entra

The process to configure SQL Server with Azure Arc for Microsoft Entra is the same as it was when we launched SQL Server 2022 with some improvements.

Let's say you would like to try out how to log in to SQL Server with Microsoft Entra and use Multi-factor Authentication. A simple way to do this is to add a Microsoft Entra admin to your SQL Server. This is very similar to how you might do this for Azure SQL.

Prerequisites

To set up Microsoft Entra, we have an excellent documentation tutorial that you can follow step by step at `https://learn.microsoft.com/sql/sql-server/azure-arc/entra-authentication-setup-tutorial`.

One nice recent improvement is to use the Azure Portal to set up a Microsoft Entra admin for my SQL Server, which automatically configures a certificate with Azure Key and Entra application. These instructions can be found at `https://learn.microsoft.com/sql/relational-databases/security/authentication-access/azure-ad-authentication-sql-server-automation-setup-tutorial`.

You are going to need some important Azure resources to go through these steps:

- Create a new Azure Key Vault (AKV). Get started at **aka.ms/azurekeyvault**.
- You will need at minimum Contributor rights in your subscription.

Be sure to follow these steps precisely as any missed step could result in a problem when configuring Microsoft Entra with Azure Arc and SQL Server.

Tip If you are an owner of any of the resources needed for these prerequisites, you do not have to add yourself as a Contributor.

When you are done, you can check the following to ensure Microsoft Entra is ready to use. First, in the Azure Portal you can see Microsoft Entra successfully configured like in Figure 6-13.

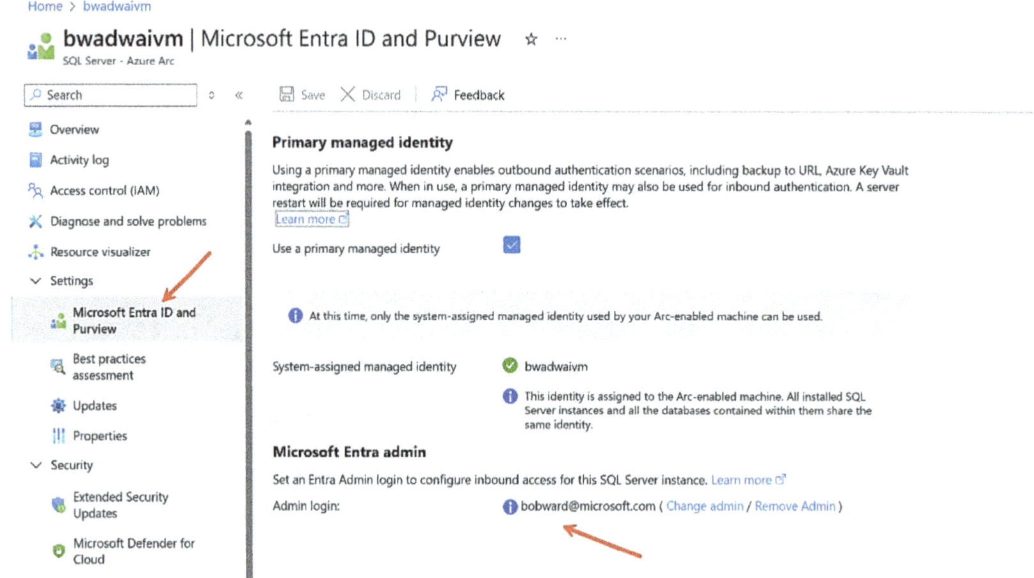

***Figure 6-13.** Microsoft Entra successfully configured*

Note Purview access policies have been discontinued for SQL Server 2025, so that option will not appear by GA.

You can also verify you are properly set up as an admin for SQL Server (e.g., sysadmin role) by trying to log in with a tool like SSMS. In my case, Microsoft requires MFA, so my login looks like Figure 6-14 (using the new connection dialog).

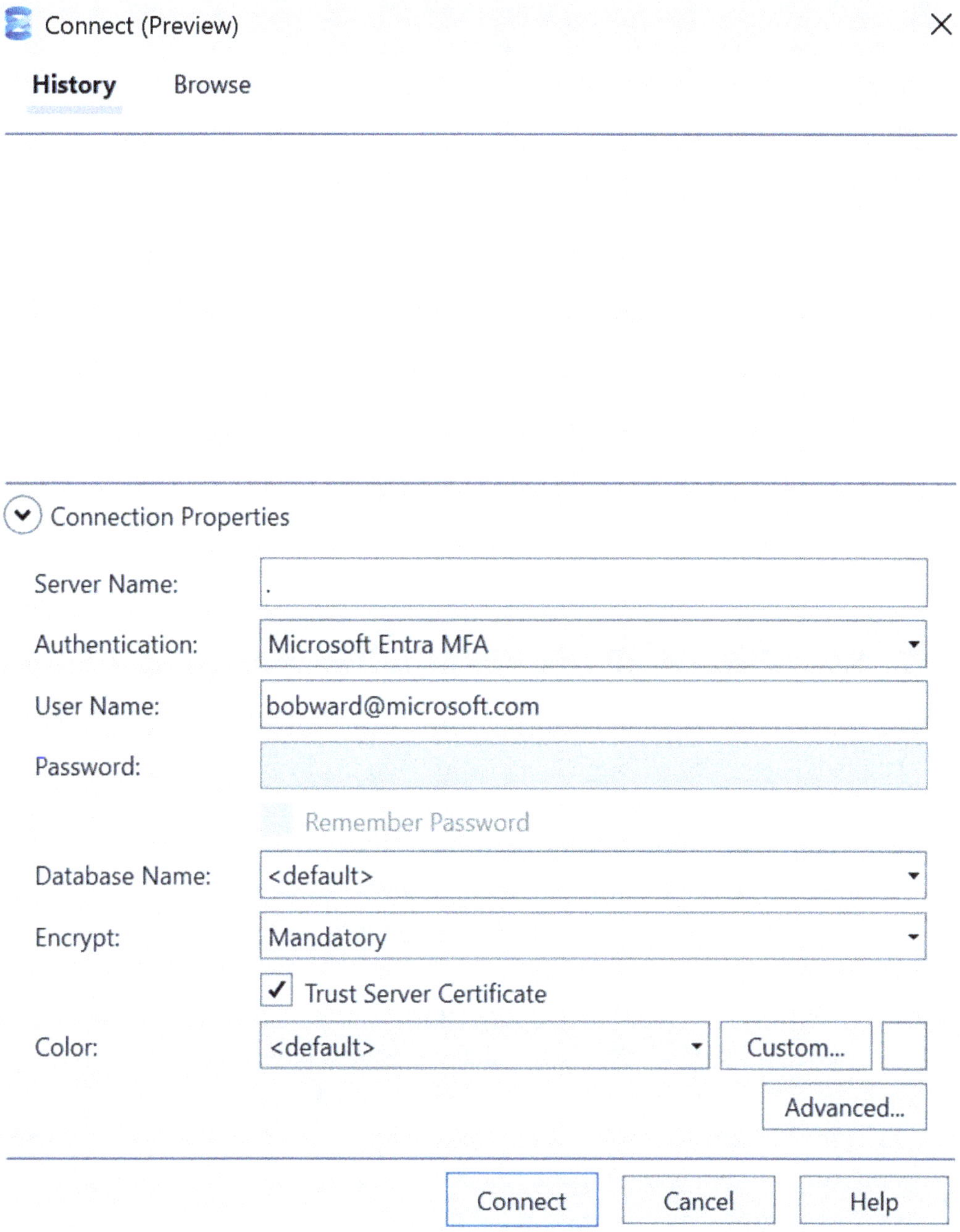

Connect (Preview)

History Browse

Connection Properties

Server Name: .

Authentication: Microsoft Entra MFA

User Name: bobward@microsoft.com

Password:

Remember Password

Database Name: <default>

Encrypt: Mandatory

Trust Server Certificate

Color: <default> Custom...

Advanced...

Connect Cancel Help

***Figure 6-14.** Logging in with SSMS using Microsoft Entra MFA*

At this point, you can add more logins and/or users with Microsoft Entra accounts and assign them with permissions as you would with SQL or Windows Domain users. A new syntax exists with an option FROM EXTERNAL PROVIDER. Learn more at `https://learn.microsoft.com/sql/sql-server/azure-arc/entra-authentication-setup-tutorial#create-logins-and-users`. This is the same syntax as with Azure SQL showing the consistency for Entra across SQL.

Tip One of the nice features of Entra I really like is the ability to create users in a database without having to create a login!

Using Microsoft Entra is a very nice feature that Arc allows for SQL Server. However, what would be interesting is to use a concept called a Managed Identity to go completely *passwordless*. Furthermore, it would be great to use this concept not just for authentication *inbound* (login) but also *outbound* out of SQL Server to Azure.

Go *Passwordless* with Microsoft Entra Managed Identity

You saw in Chapter 4 of the book how to build an AI application with vector searching using AI models such as those found in Azure AI Foundry. In this example you used the concept of an API key to access the AI model endpoint. This is akin to using a password string, which is clear text. What if there was a better way?

Along comes a Microsoft Entra Managed Identity. A Managed Identity is a security object enabled and authenticated by Microsoft Entra. A Managed Identity comes in two flavors: system-assigned or user-assigned. A system-assigned Managed Identity is associated with an Azure resource and has the same lifecycle of an Azure source. A user-assigned Managed Identity is a standalone security object that can be used across multiple Azure resources. For SQL Server enabled by Azure Arc, you can use a system-assigned Managed Identity. A Managed Identity has no certificate or password. Applications that are authenticating with the resource assigned to the identity can request a token from Microsoft Entra and use this token to gain access to an Azure resource that has also been assigned to the same identity.

Figure 6-15 shows how this works from a visual perspective to use a system-assigned Managed Identity to access an AI model in Azure AI Foundry.

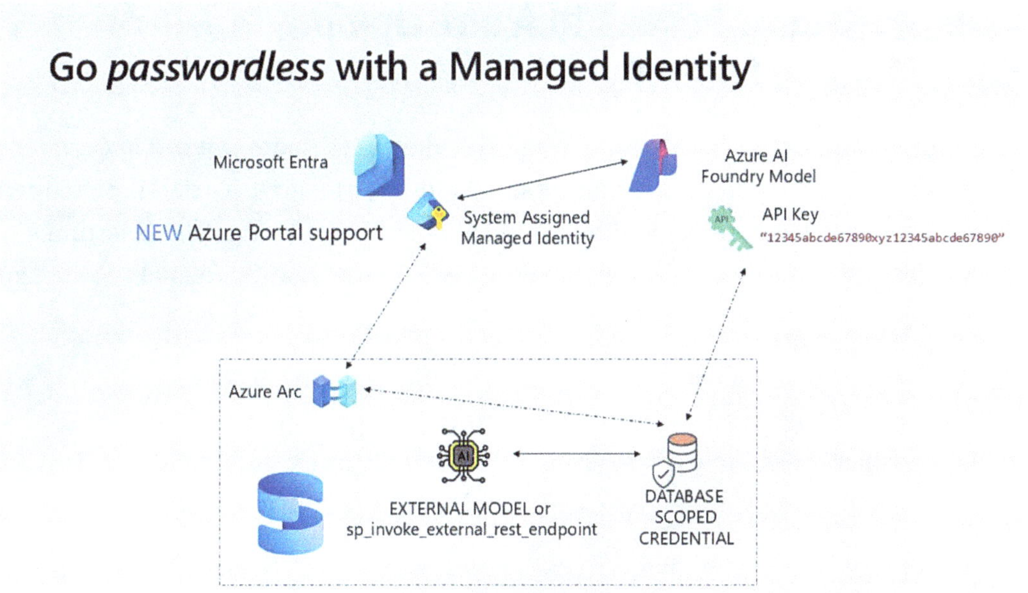

Go *passwordless* with a Managed Identity

Microsoft Entra
NEW Azure Portal support
System Assigned
Managed Identity
Azure AI
Foundry Model
API Key
"12345abcde67890xyz12345abcde67890"
Azure Arc
EXTERNAL MODEL or
sp_invoke_external_rest_endpoint
DATABASE
SCOPED
CREDENTIAL

Figure 6-15. *Using a Managed Identity to access Azure AI*

In Chapter 4 you assigned the API key associated with the Azure AI model to a **database scoped credential**. Now we can use the system-assigned Managed Identity that is associated with the Azure Arc resource. The Azure Portal experience you used previously in this chapter also configured the system-assigned Managed Identity to be associated with the SQL Server instance.

You can use the system-assigned Microsoft Entra Managed Identity to both log in to SQL Server (inbound) and to authenticate (outbound) to Azure resources such as Azure Storage, Azure Key Vault, and Azure AI.

Let's look at the example you used in Chapter 4, but this time you will use the Managed Identity to authenticate the Azure AI model instead of the API key.

Tip You can see if SQL Server recognizes a Managed Identity using the Dynamic Management View **sys.dm_server_managed_identities**.

Example: Secure Access to Azure OpenAI with a Managed Identity

This example assumes you set up the example to connect to Azure OpenAI in Chapter 4 of the book. I will show you how to make a small modification to use a database scoped credential to use the Managed Identity instead of the API key. This example assumes you have already gone through the previous example in this chapter to enable Microsoft Entra for SQL Server using Azure Arc.

1. In order to use a database scoped credential with a Managed Identity, you need to use a server configuration option that enables the use of the system-assigned identity by executing the following T-SQL (or use the script **enablemidbscopecred.sql**):

   ```
   EXECUTE sp_configure 'allow server scoped db credentials', 1;
   RECONFIGURE WITH OVERRIDE;
   ```

2. You must now assign permissions to your Azure OpenAI resource for the system-assigned Managed Identity from the Azure Arc resource. The steps can be found in this documentation page: `https://learn.microsoft.com/azure/ai-foundry/openai/how-to/role-based-access-control`. For me, the Azure OpenAI service is called productsopenai. I found this service in the Azure Portal. Then I used the option called Access control (IAM) to add a role assignment. I assigned the **Cognitive Services OpenAI Contributor** role to the system-assigned Managed Identity for the Azure Arc resource. When you select this, it will be the name of the VM and identified as Machine - Azure Arc.

3. With this now set up, we can go back and create an updated database scoped credential to use the Managed Identity. Since the external model is still using our database scoped credential, that must be dropped first. Use the script **credsusingmi.sql** that has the T-SQL statements:

   ```
   USE AdventureWorks;
   GO
   DROP EXTERNAL MODEL MyAzureOpenAIEmbeddingModel;
   ```

```
GO
DROP DATABASE SCOPED CREDENTIAL [https://<azureaiservice>.openai.
azure.com];
GO
-- Create access credentials to Azure OpenAI using a managed
identity:
CREATE DATABASE SCOPED CREDENTIAL [https://<azureaiservice>.
openai.azure.com]
    WITH IDENTITY = 'Managed Identity', secret =
    '{"resourceid":"https://cognitiveservices.azure.com"}';
GO
-- Create the EXTERNAL MODEL
CREATE EXTERNAL MODEL MyAzureOpenAIEmbeddingModel
WITH (
      LOCATION = 'https://<azureariservice>.openai.azure.com/
      openai/deployments/text-embedding-ada-002/embeddings?api-
      version=2023-05-15',
      API_FORMAT = 'Azure OpenAI',
      MODEL_TYPE = EMBEDDINGS,
      MODEL = 'text-embedding-ada-002',
      CREDENTIAL = [https://<azureaiservice>.openai.azure.com]
);
GO
```

You need to substitute in your <azureaiservice> with the name of the Azure OpenAI service you deployed in Chapter 4.

4. There is no need to rebuild your embeddings. An easy way to test to ensure all still works is run the prompt test as you did in Chapter 4 of the book using the script **find_products_prompt_vector_search.sql**. This is because the T-SQL function **AI_GENERATE_EMBEDDINGS**() is used to take a prompt in a stored procedure and use the defined model to generate an embedding. This would fail if you had not set up the new scoped credential correctly.

If you want to try other examples with Microsoft Entra Managed Identity and SQL Server 2025, check out the new security features for SQL Server 2025 at `https://learn.microsoft.com/sql/sql-server/what-s-new-in-sql-server-2025#security`.

Connecting SQL to the World

SQL Server is now not just connected to Azure. SQL Server is *integrated* with Azure to bring to you powerful solutions such as managed disaster recovery, near-real time analytics, and centralized security and governance. Each of these solutions is required to enhance the SQL Server engine so that the capabilities you need are seamless and work with the SQL Server ecosystem. There is no requirement to use all of these services. Pick and choose the one you want when you need it. The future for SQL Server hybrid is bright, and for Microsoft and our customers, the journey is just starting.

Travis Wright, Partner Group Engineering Manager at Microsoft, summed up the value of Azure Arc very well, *"By enabling SQL Server with Azure Arc, customers gain powerful benefits by extending the management capabilities of Azure that bridge their on-premises or multi-cloud environments. By simply installing the Azure Arc agent on a machine, the details about the OS, hardware, SQL Server instances, databases, and availability groups are published to Azure to create an always-up-to-date inventory management system securely accessible from anywhere. Operations are streamlined with features like automated patching, backups and point-in-time restore, advanced performance monitoring, and best practices assessments—all accessible through the Azure Portal. Security can be greatly enhanced by using Entra ID authentication and managed identities for secretless, hassle-free inbound apps and user authentication to SQL Server and for outbound connections to Azure services like blob storage for backups. Customers can save money and simplify licensing by switching to the pay-as-you-go model billed through Azure and then scaling up/down as needed. When customers are ready, they can use the built-in automated migration assessments to get recommendations on the ideal target for running SQL Server in VMs in Azure or in Azure SQL PaaS and then set up replication or log shipping automatically to begin the migration."*

CHAPTER 7

The Core Engine of SQL Server 2025

I vividly remember going into a meeting in the late summer of 2024 to review all the new features of SQL Server 2025. Our team had put such energy into AI as a focus for this release I honestly was a bit concerned about what we might have sacrificed regarding the core SQL Server engine, specifically features for security, performance, and availability. I knew some of the major engine features we could pull from the cloud like optimized locking (OL), but I was not sure how many features could make our release schedule.

I walked away from the meeting not just relieved but excited. I saw in our roadmap across our preview builds so many features for the engine with real value. This is because we had many features already baking in the cloud and in our "backlog" of work that didn't make the SQL Server 2022 release. It was also a testament to the determination and hard work for many engineers and product managers to bring in quality features based on customer feedback with true value.

In this chapter I'll give you an overview of the major engine features and then dive into categories of security, performance, and availability. Then I will finish off the chapter with a few "hidden gems." Throughout the chapter I'll have examples for you to use, and here is the good news. All you need is a Developer Editon of SQL Server 2025 to do any of the examples. No large datasets are required, so all of this can be done on your laptop! Some of these examples are "inline" to show a feature. Some are "full examples" with scripts. All scripts for these can be found in the **ch7 – Core Engine** folder for examples for the book.

This is a very long chapter, so you may want to consider jumping to the section of your choice: "Security," "Performance," "Availability," or "Hidden Gems." Or just plow through it all from the beginning. I also have a fun section for you at the end of the chapter talking about how we make SQL Server 2025 the fastest database on the planet.

B. Ward, *SQL Server 2025 Unveiled*, https://doi.org/10.1007/979-8-8688-1847-9_7

What's New for the Engine?

As I mentioned in Chapter 5 of the book, all the T-SQL enhancements and developer-oriented features that are part of the engine like JSON are covered in that chapter. This chapter is more focused on core engine capabilities specific to security, performance, and availability.

For **security**, there are new features ranging from *passwordless* authentication with Managed Identities, security cache improvements, and enhancements for encryption and password security.

Performance features can be classified into three categories: (1) features designed to accelerate query performance; (2) features designed to improve concurrency, which implicitly can help speed up overall performance; and (3) features designed to help manage query performance.

Availability is really about High Availability and Disaster Recovery (HADR). This includes features designed to enhance Always On Availability Groups (AGs) and backups.

I built a slide for my presentations that show how vast our engine investments have been for SQL Server 2025 as seen in Figure 7-1.

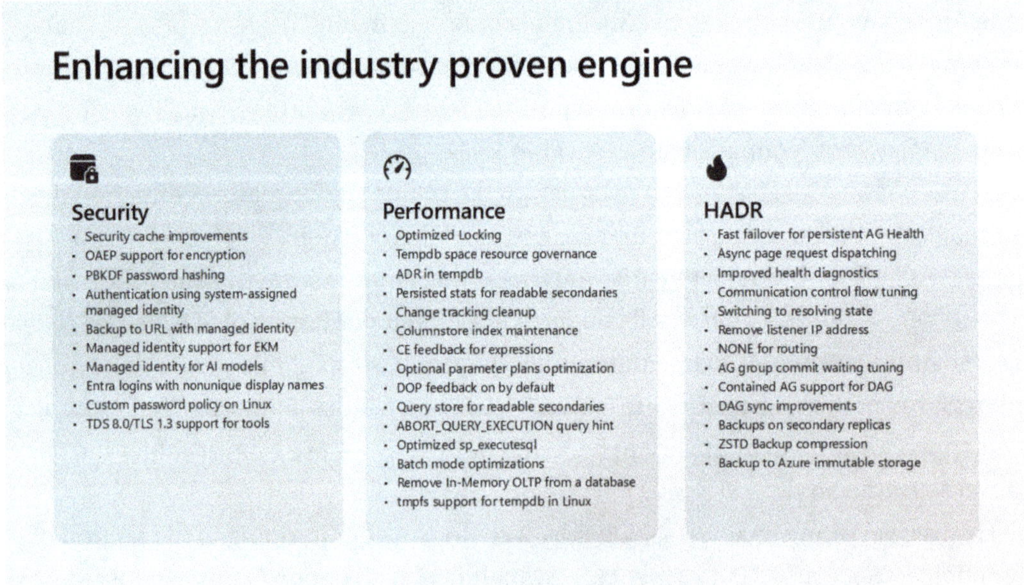

Figure 7-1. *SQL Server 2025 core engine investments*

Let's start by looking closer at security for SQL Server 2025.

Security

Security enhancements include a range of new capabilities including authentication, encryption, password protection, and concurrency.

Microsoft Entra and Managed Identity

You read extensively in Chapter 6 of the book about Microsoft Entra authentication including *passwordless* with a Managed Identity. Entra is a managed cloud service for security and enabled for SQL Server through Azure Arc. In Chapter 6 you learned how to connect with a Microsoft Entra account and to secure access to Azure AI Foundry models with a Managed Identity. There are a few other scenarios enabled by Entra worth looking at including BACKUP and Azure Key Vault. I'll also mention using a Managed Identity with PolyBase in the last section of the chapter called "Hidden Gems."

BACKUP to URL with Managed Identity

SQL Server has supported backing up databases to Azure Blob Storage for many releases. However, to back up a database to Azure Blob Storage, you must create a credential in SQL Server that uses an access key or Shared Access Signature (SAS) token from the storage account. This is similar to the API key I showed you in Chapter 4 of the book for access to AI models.

In order to provide a more secure method, you can use a Managed Identity, similar to how you configured a Managed Identity to access an AI model in Chapter 6 of the book.

The process is simple. You will assign the system-assigned Managed Identity from the Azure Arc resource permissions to the Azure Storage account. Then you will create a SQL Server credential associated with the Managed Identity, much like you did for the database scoped credential in Chapter 6. Now you should have permission to execute a BACKUP to URL (or RESTORE from URL) for your database just as you would with an access key or SAS token. Read all the details at `https://learn.microsoft.com/sql/sql-server/azure-arc/backup-to-url`.

Managed Identity Support for Azure Key Vault

Traditionally, keys used for encryption features like Transparent Data Encryption (TDE) are generated and stored in SQL Server. In SQL Server 2016, we added the capabilities to generate and store these keys in Azure Key Vault (AKV) as you can see in Figure 7-2.

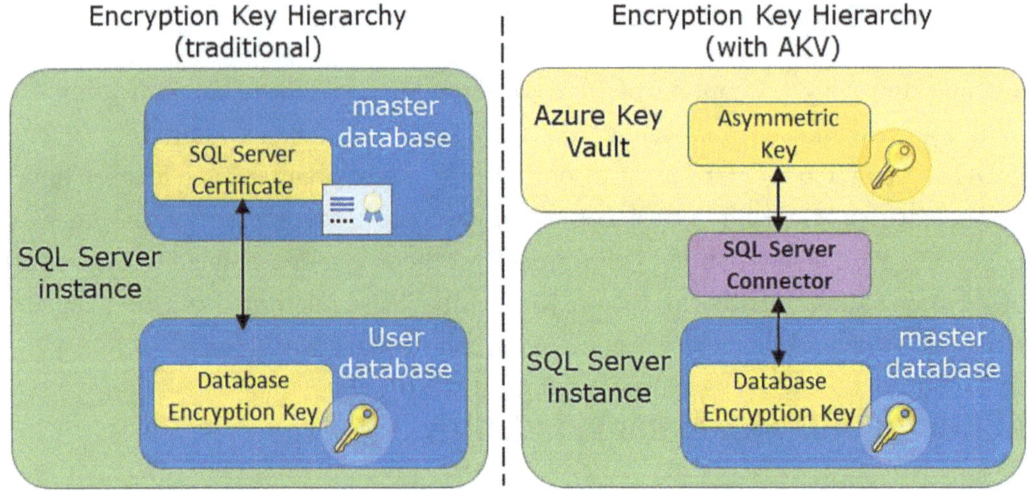

Figure 7-2. *Techniques to store keys for encryption in SQL Server*

The architecture on the right uses a concept called Extensible Key Management (EKM). With this concept you must create a service principal in Microsoft Entra as a client secret (a.k.a. password) and use that secret and a credential in SQL Server. Do you see the pattern here? Everything feels like a password.

Now with SQL Server 2025, you can assign the system-assigned Managed Identity for the Azure Arc resource associated with SQL Server 2025 permissions to Azure Key Vault. Then you can create the credential required for EKM with the Managed Identity. Learn more at `https://learn.microsoft.com/sql/sql-server/azure-arc/managed-identity-extensible-key-management`.

Microsoft Entra Logins for Service Principals

Microsoft Entra with Azure Arc does allow for a service principal to be used as a login.

> **Note** Not to be confused with creating a login with the *system-assigned Managed Identity*, which was described in Chapter 6.

One of the problems with using a service principal, also known as an application, is the application name, or display name, doesn't have to be unique in Microsoft Entra. Now with SQL Server 2025, you can use an extension to the **CREATE LOGIN** statement **WITH OBJECT_ID** to specify the object id of the service principal. Learn more at `https://learn.microsoft.com/azure/azure-sql/database/authentication-microsoft-entra-create-users-with-nonunique-names`.

Security Cache

If possible, we cache *everything* in SQL Server. Anything cached is always faster than reading something from disk (well, ok, most of the time). SQL Server has a sophisticated security system including many different types of *securables* and permissions. Therefore, to ensure proper permissions are checked for everything, we must look up information in our security system tables. The security cache of SQL Server is effectively a cache of the security system tables for the instance and databases. A common name for a large part of this cache is called the **TokenAndPermUserStore**.

When SQL Server is started, nothing is in the security cache. As logins connect or other security operations occur, the cache gets populated. Some scenarios require parts of or the entire cache to be *invalidated*. Memory pressure is one of these, but certain security management operations can also cause cache invalidation. Invalidation can slow down subsequent connections or queries in order to look up information from disk and populate the cache.

One scenario that is **not** optimized is security management operations for logins associated with the master database (or instance). These operations can invalidate the cache for *all logins*, even if the operation is specific to a certain login.

Pieter Vanhove, Senior Product Manager for SQL security, showed me one such scenario.

> *"Let's say you have two logins, Login1 and Login2, and you have connected with these logins so there are cached entries for each. If you come along and perform an operation against Login2, like grant it permissions, the cache is invalidated for Login1 even though it was not affected by the operation. You may be wondering why this matters? This is because any cache invalidation in master effectively invalidates the entire security cache. All logins and databases are affected!"*

In SQL Server 2025, this behavior is now enhanced so only the cache for Login2 would be affected. This feature applies to the CREATE, ALTER, and DROP login scenarios and permission changes for individual logins. Group logins continue to experience server-level invalidation. You can read more about this enhancement at `https://learn.microsoft.com/sql/relational-databases/security-cache#known-issues`. This documentation also lists what operations cause various cache invalidations including the scope (server or database).

Note It is important to know that this enhancement does not affect performance scenarios with a *large* TokenAndPermUserStore. The performance scenarios described in `https://learn.microsoft.com/troubleshoot/sql/database-engine/performance/token-and-perm-user-store-perf-issue` could still be something for you to consider especially on systems with large amounts of memory.

Encryption and Password Enhancements

There are several enhancements in SQL Server 2025 to support the latest encryption and password requirements of modern applications.

TLS 1.3/TDS 8.0

In SQL Server 2022, we introduced a new version of our Tabular Data Stream (TDS) protocol, version 8.0. TDS 8.0 changes the method in which network packets are processed for encryption for the initial connection to SQL Server (called a *handshake*). Now, the Transport Layer Security (TLS) precedes any TDS messages, wrapping the TDS session in TLS to enforce encryption, making TDS 8.0 aligned with HTTPS and other web protocols.

This new version allows SQL Server to support clients that use the latest TLS 1.3 version. To use TDS 8.0, clients must use the *strict* connected encryption (see `https://learn.microsoft.com/sql/relational-databases/security/networking/tds-8#strict-connection-encryption`) option when connecting to SQL Server.

This is a much more secure method for client applications to connect to SQL Server. The problem? Our own tools and features were not updated in SQL Server 2022 to use this new secure method.

In SQL Server 2025, several tools and features now support the use of TDS 8.0/ TLS 1.3. This list is extensive and includes everything from sqlcmd.exe all the way to Linked Servers.

I should stop here to ensure you know that some of these changes could cause an issue for you when you upgrade or use SQL Server 2025. Take, for example, Linked Servers. In SQL Server 2025, the default provider, which is the Microsoft OLEDB 19 driver, which is required to support TLS 1.3 and TDS 8.0, will **require you to use encryption by default**. You always have the option to use the OLEDB 18 driver or change the connection strings to not use strict encryption.

Caution One important issue I think you might encounter is with Linked Servers. A "SQL Server" Linked Server will now use OLEDB 19 and require encryption by default. If you don't specify the option to trust a server certificate, new Linked Servers may fail to be created by default. We are looking to optimize the SSMS experience to help you if you use that tool to create new Linked Servers.

You can see a list of features that you might encounter issues with at `https://learn.microsoft.com/sql/database-engine/breaking-changes-to-database-engine-features-in-sql-server-2025`.

Keep up with the latest list of all features and tools that now support TDS 8.0 and TLS 1.3 at `https://learn.microsoft.com/en-us/sql/relational-databases/security/networking/tds-8#sql-server-2025-support`.

OAEP Padding Mode Support for RSA Encryption

Our security team is always looking to make sure we use the latest security technology for features like encryption. When SQL Server uses encryption for things like creating an asymmetric key, it uses encryption techniques like Rivest–Shamir–Adleman (RSA). This technique is common in the industry but isn't strong enough on its own, so a *padding* technique is used called Public-Key Cryptography Standards (PKCS) padding. PKCS version 1.5 has been used for some time but also has some weaknesses that have been discovered. Therefore, the latest version of PKCS, version 2.X, supports a concept called Optimal Asymmetric Encryption Padding (OAEP). SQL Server 2025 now uses this technique. This feature requires dbcompat 170, and you will need to create a new key, which will use this new technique.

Note For SQL Server, we rely on the latest version of the Windows Cryptography API to support this. Linux crypt libraries provide the same capabilities.

PBKDF for Password Hashes

We understand the reality that SQL authentication is still used by our customers, which includes the use of passwords. So, while we want our customers to look for new authentication systems like Microsoft Entra, we want to make sure we secure your passwords as best as we can.

When SQL Server stores a clear-text password, say from a CREATE LOGIN statement, we don't store the clear text. We use algorithms like hashing to generate a value that can't be reverse-engineered (in a way like an encryption).

In SQL Server 2025, we have brought in an enhanced algorithm for hashing the passwords known as RFC2898, also known as a *password-based key derivation function* (PBKDF). This technique increases the number of times we hash the password, increasing security. If you really want to dive into the security details of this, you can read `https://www.ietf.org/rfc/rfc2898.txt`. One benefit of this new method is it complies with the NIST SP 800-63b guidelines (`https://pages.nist.gov/800-63-3/sp800-63b.html`).

When you create a new login or alter it, we will use the new hash algorithm. Interesting enough, when you log in to SQL Server 2025, we can recognize if the hashed password stored for the login uses the new algorithm and, if not, convert it. This would occur for any in-place upgrades from previous versions of SQL Server or if you scripted out logins from older versions (scripting logins build a password with a random value hashed). If you observe the time it takes to log in, you might see a slight delay (in ms) because of the new algorithm.

Custom Password Policy for SQL Server on Linux

For a very long time, SQL Server has allowed you to configure specific policies for your passwords for SQL Server authentication, for example, password complexity. You can see all the ways to set these policies and enforce them on Windows at `https://learn.microsoft.com/sql/relational-databases/security/password-policy`.

SQL Server on Linux did not have these options. Starting in SQL Server 2025, you can use the mssql.conf file and adutil utility to enforce these same types of password policies. Read more at `https://learn.microsoft.com/sql/linux/sql-server-linux-custom-password-policy`.

Performance

There is no question security and availability for mission-critical SQL Server applications are *table stakes*. While having consistent and great performance at scale is very important, there are levels of this that are acceptable for SQL Server deployments depending on the requirements of the application and business. And performance is just fun (well, it is at least for me) and quite frankly a bit easier to show by example.

Therefore, this section of the chapter has the most examples of any of the sections. Each example in this section will list the prerequisites. All examples will require you to install SQL Server 2025. Developer Edition (Enterprise Version) will work well to try these features. I also use SQL Server Management Studio (SSMS), but you are free to use any tool that can run SQL statements or scripts.

For performance, I've broken down this section into important features for application availability, tempdb enhancements, query optimization and execution, and query management.

Let's start by looking at two significant features in this release that help improve performance because they help in improving application availability and concurrency: **optimized locking** and **tempdb resource governance**.

Optimized Locking

I've seen developers for years struggling with having to know the internals of locking in SQL Server in order to meet the concurrency needs of their application. And most of these scenarios revolve around making changes to a table for a certain set of rows where logically two different sets of changes "shouldn't affect each other." Consider a scenario where a query updates 10,000 rows that is based on cluster indexing ordering and these rows are at the "front of the table." And then another update is executed for a single row at the "end of the table." As a developer, you would naturally expect these updates to not block each other. But unfortunately, there is a pesky problem called *lock escalation* that can occur and cause these two updates to block. And there are other scenarios where even updates to a single row logically shouldn't block each other, but because of the way we implement locks, blocking can occur. Developers are left scratching their heads wondering how to write code in a way to avoid these problems.

This is why we have introduced the concept of *optimized locking*. As I talk more about how it works, you can follow along as well in the documentation at `https://learn.microsoft.com/sql/relational-databases/performance/optimized-locking`.

Note We have been perfecting this technology for some time in Azure. I gave a sneak peek of optimized locking with Conor Cunningham at the PASS Summit 2022 keynote!

How It Works

There are two different aspects for optimized locking: (1) elimination of lock escalation through *transaction* locks and (2) Lock After Qualification (LAQ). LAQ builds on the transaction lock concept.

We introduced a new feature in SQL Server 2019 called **Accelerated Database Recovery (ADR)**. The intention of this feature was to provide faster rollback and recovery of databases by introducing row versioning *inside* the database. One of the artifacts of ADR is that each "version" of the row on a page has a **Transaction ID** (TID) associated with the transaction that modified the row.

Note You can see all the details of how row versions work with ADR at the original whitepaper from `https://aka.ms/sqladr`.

So we discovered if you have ADR enabled, we could use the TID as a locking mechanism avoiding having to keep row locks for every row the transaction is modifying. We would still need to acquire the necessary page and row lock as each row is modified, but they can be released after the modification only keeping the TID lock until the transaction is committed or rolled back. This means now for a large update, we *don't have to accumulate* row locks. Why is this important and what does this have to do with lock escalation? This is because SQL Server by default will escalate to a higher-level lock if too many locks are acquired by a transaction. This can become a major problem because in most cases this results in the acquisition of a table lock. The thinking around lock escalation is the overhead of holding on to so many locks (which require resources like memory) vs. just using one lock.

Lock escalation has been around since, like, forever, and there are several ways to try and get around it (you can read about the details on thresholds and how they work at `https://learn.microsoft.com/sql/relational-databases/sql-server-transaction-locking-and-row-versioning-guide#lock-escalation`).

Now using the power of row versioning, we can simply acquire and hold an exclusive lock on the TID instead of accumulating row locks. We call these **XACT** locks. Using an XACT lock allows us to protect only a transaction related to *versions* of rows. Other queries may need to acquire a shared lock on the TID, but only if you care about row version affected by that transaction. The result is pure beauty. **No more lock escalation problems!**

The basic requirement to enable this feature is to use **ALTER DATABASE** to set **ACCELERATED_DATABASE_RECOVERY** and **OPTIMIZED_LOCKING** ON. OPTIMIZED_LOCKING can only be set if ACCELERATED_DATABASE_RECOVERY is ON.

Before you read on, you may not be using ADR and wonder what the ramifications are to use it and how it might affect your application. First, with any option like this, you must always test everything thoroughly. But to give you some solace, ADR is on by default in Azure SQL and has been for several years. In addition, you can read more about any performance impact or storage overhead in our documentation at `https://learn.microsoft.com/sql/relational-databases/accelerated-database-recovery-concepts` and our whitepaper at `https://aka.ms/sqladr`. We also made some enhancements in SQL Server 2022 based on customer feedback. I honestly have a mission now to see if I can get all our SQL Server customers to use ADR.

Note I often get asked why the default on-premises isn't ON for ADR. That is because we take a more conservative approach to make a setting a default given how many different ways customers can deploy and configure SQL Server. But maybe one day we can do this for ADR.

So now that you have eliminated lock escalation problems, is there anything else optimized locking can help you with? The answer is yes. ADR and OPTIMIZED_LOCKING together are independent of **transaction isolation levels**. No matter what isolation level you use, we can use XACT locks to avoid lock escalation problems. What about a scenario where a transaction needs to determine whether a row qualifies for an update? SQL Server will typically attempt to acquire locks *before* updating a row, for example, when using the typical default of the READ_COMMITTED transaction isolation level. So even with XACT locks, one update may get blocked by another because the first update is keeping locks to uphold transaction isolation levels.

When you enable the READ_COMMITTED_SNAPSHOT option for your database, along with ADR and OPTIMIZED_LOCKING, you now allow SQL Server to use the READ_COMMITTED_SNAPSHOT isolation level *combined with XACT locks*. Now, SQL Server will optimistically only acquire the locks it needs **after** determining the row it needs to update. This is the concept of **Lock After Qualification (LAQ).** Think of this as a "lock-free predicate evaluation."

There are some caveats of LAQ including scenarios where predicates need to be re-evaluated. Be sure to read more details in the documentation about these scenarios and how to diagnose any issues with optimized locking at `https://learn.microsoft.com/sql/relational-databases/performance/optimized-locking#diagnostic-additions-for-optimized-locking`.

In summary, to enable the full power of optimized locking, you need these database options enabled:

- ACCELERATED_DATABASE_RECOVERY
- OPTIMIZED_LOCKING
- READ_COMMITTED_SNAPSHOT

Note ACCELERATED_DATABASE_RECOVERY and OPTIMIZED_LOCKING are on by default for Azure SQL and cannot be disabled. READ_COMMITED_SNAPHOT is on by default but can be disabled.

There is no better way to get the full picture of how this works than an example.

Prerequisites for the Example

You can find all scripts in the samples with the book in **the ch7 – Core Engine\optimized_locking** folder. I use SQL Server Management Studio (SSMS) for all script and T-SQL statement execution.

- Download the database sample **AdventureWorks** from `https://github.com/Microsoft/sql-server-samples/releases/download/adventureworks/AdventureWorks2022.bak`.
- Restore the database using the script **restore_adventureworks.sql** (you may need to edit the file paths for the backup and/or database and log files).
- Enable Accelerated Database Recovery for the database AdventureWorks using the script **enableadr.sql**.
- Make sure optimized locking is *disabled* by executing the script **disableoptimizedlocking.sql** in case it is enabled**.** We will enable ADR but disable optimized locking so you can see lock escalation but make it easier to enable it later.

Example Steps for Lock Escalation

Let's first see how to show how optimized locking can help with lock escalations.

Example 1: Show lock escalations

1. Load the script **getlocks.sql**. You will use this script to observe locking behavior. This script executes the following T-SQL statement:

```
SELECT resource_type, resource_database_id, resource_associated_
entity_id, -- resource_description, request_mode, request_session_
id, request_status, COUNT(*) AS lock_count
FROM sys.dm_tran_locks
WHERE resource_type != 'DATABASE'
GROUP BY resource_type, resource_database_id, resource_associated_
entity_id, request_mode, request_session_id, request_status
ORDER BY resource_type, resource_database_id, resource_associated_
entity_id, request_mode, request_session_id, request_status;
GO
```

2. Load the script **updatefreightsmall.sql** in a SSMS query editor window. This script runs the following T-SQL statements:

```
USE AdventureWorks;
GO
-- Run this batch first to update 2500 rows
DECLARE @minsalesorderid INT;
SELECT @minsalesorderid = MIN(SalesOrderID) FROM Sales.
SalesOrderHeader;
BEGIN TRAN
UPDATE Sales.SalesOrderHeader
SET Freight = Freight * .10
WHERE SalesOrderID <= @minsalesorderid + 2500;
GO
-- Rollback the transaction when needed
ROLLBACK TRAN;
GO
```

   This script will increase the freight costs for each order by 1 for the first 2,500 rows.

   Execute the first batch in the query script up to the GO statement**. Do not execute the ROLLBACK TRAN statement.**

3. Switch to the query editor window for **getlocks.sql** and execute the query. Your results should look like Figure 7-3.

	resource_type	resource_database_id	resource_associated_entity_id	request_mode	request_session_id	request_status	lock_count
1	KEY	5	72057594051166208	X	65	GRANT	2501
2	METADATA	5	0	Sch-S	65	GRANT	1
3	OBJECT	5	1602104748	IX	65	GRANT	1
4	PAGE	5	72057594051166208	IX	65	GRANT	111

***Figure 7-3.** Row locks accumulated without optimized locking*

   You can see SQL Server has accumulated ~2,500 row locks for the transaction. If the number of row locks in an active transaction gets too big, lock escalation can kick in. Let's see what that looks like.

4. First, execute the **ROLLBACK TRAN** statement in the **updatefreightsmall.sql** script. This will release all locks and allow the next step to proceed. You can close out this query.

5. Load the script **updatefreightbig.sql** that runs the following T-SQL statements:

```
USE AdventureWorks;
GO
-- Run this batch first to update 10000 rows
DECLARE @minsalesorderid INT;
SELECT @minsalesorderid = MIN(SalesOrderID) FROM Sales.
SalesOrderHeader;
BEGIN TRAN
UPDATE Sales.SalesOrderHeader
SET Freight = Freight * .10
WHERE SalesOrderID <= @minsalesorderid + 10000;
GO

-- Rollback the transaction when needed
ROLLBACK TRAN;
GO
```

   This script will increase the freight costs for each order by 1 for the first 10,000 rows. **Execute the first batch in the query script up to the GO statement. Do not execute the ROLLBACK TRAN statement.**

6. Switch to the query editor window for **getlocks.sql** and look at the results like Figure 7-4.

	resource_type	resource_database_id	resource_associated_entity_id	request_mode	request_session_id	request_status	lock_count
1	METADATA	5	0	Sch-S	65	GRANT	1
2	OBJECT	1	28	Sch-S	40	GRANT	1
3	OBJECT	5	1602104748	X	65	GRANT	20

Figure 7-4. *Locks after lock escalation*

You can see from these results that an exclusive object (or table) lock has occurred. This is what lock escalation looks like.

7. Load the script **updatemaxfreight.sql** in a SSMS query editor window that runs the following T-SQL statements:

```
USE AdventureWorks;
GO
-- Update the highest salesorderid
DECLARE @maxsalesorderid INT;
SELECT @maxsalesorderid = MAX(SalesOrderID) FROM Sales.
SalesOrderHeader;
BEGIN TRAN
UPDATE Sales.SalesOrderHeader
SET Freight = Freight * .10
WHERE SalesOrderID = @maxsalesorderid;
GO
-- Rollback the transaction when needed
ROLLBACK TRAN;
GO
```

This script updates the freight for a row *not affected* by the previous update (or shouldn't be). **Execute the first batch in the query script up to the GO statement. Do not execute the ROLLBACK TRAN statement.** Notice this batch does not complete. This is because the update is blocked by the OBJECT X lock.

8. Execute the query in **getlocks.sql** again. Your results should look like Figure 7-5.

	resource_type	resource_database_id	resource_associated_entity_id	request_mode	request_session_id	request_status	lock_count
1	METADATA	5	0	Sch-S	65	GRANT	1
2	OBJECT	5	1602104748	IX	73	WAIT	1
3	OBJECT	5	1602104748	X	65	GRANT	20

Figure 7-5. *Blocking due to lock escalation*

You can see from the results that the update of just one row (the max value in the table) is blocked immediately because it needs an intent exclusive lock on the table (this is normal for even updating one row). Not good at all.

9. Clean up current transactions. Roll back the transactions in **updatefreightbig.sql** and **updatefreightmax.sql**. You can do this by executing the **ROLLBACK TRAN** statement in those scripts. Leave all the query editor windows open.

Example 2: Lock escalation eliminated after optimized locking
Now let's run the same scenario using optimized locking.

1. Enable optimized locking by loading and executing the script **enableoptimizedlocking.sql** in a SSMS query editor window. This will enable optimized locking for the database AdventureWorks.

2. Execute the T-SQL statements again (but not the ROLLBACK TRAN) in the script **updatefreightbig.sql**.

3. Execute the T-SQL statements again in **updatemaxfreight.sql** (but not the ROLLBACK TRAN). Notice this time, you are not blocked!

4. Execute the T-SQL statement in **getlocks.sql** to observe locks. Your results should look like Figure 7-6.

	resource_type	resource_database_id	resource_associated_entity_id	request_mode	request_session_id	request_status	lock_count
1	METADATA	5	0	Sch-S	65	GRANT	1
2	METADATA	5	0	Sch-S	115	GRANT	1
3	OBJECT	5	1602104748	IX	65	GRANT	1
4	OBJECT	5	1602104748	IX	115	GRANT	1
5	XACT	5	0	X	65	GRANT	1
6	XACT	5	0	X	115	GRANT	1

Figure 7-6. *Locks with no lock escalation*

So beautiful! No waits and no blocking. Each transaction has its own XACT locks. No lock escalation has occurred, and updates that should not block each other are not affected. No code changes required!

5. Clean up current transactions. Roll back the transactions in **updatefreightbig.sql** and **updatefreightmax.sql**. You can do this by executing the **ROLLBACK TRAN** statement in those scripts. Exit all scripts *except* for **getlocks.sql**.

Example Steps for Lock After Qualification (LAQ)

Let's look at the second scenario to build on transaction locks with Lock After Qualification (LAQ).

Example 1: Locking without LAQ

1. The AdventureWorks sample database has READ_COMMITTED_SNAPSHOT enabled by default, so run the script **disablercsi.sql** to disable it.

2. Load the **updatefreightpo1.sql** script. This script executes the following T-SQL statements:

```
USE AdventureWorks;
GO
-- Update a specific purchase order number
DECLARE @minsalesorderid INT;
SELECT @minsalesorderid = MIN(SalesOrderID) FROM Sales.
SalesOrderHeader;
BEGIN TRAN;
UPDATE Sales.SalesOrderHeader
SET Freight = Freight * .10
WHERE PurchaseOrderNumber = 'PO522145787';
GO

-- Rollback transaction if needed
ROLLBACK TRAN;
GO
```

   This script will update the freight for a specific PurchaseOrderNumber but requires a scan of the clustered index to find the row to update. **Execute the first batch in the query script up to the GO statement. Do not execute the ROLLBACK TRAN statement.**

3. Load the **updatefreightpo2.sql** script. This script runs the following T-SQL statements:

```
USE AdventureWorks;
GO
-- Update a specific purchase order number
DECLARE @minsalesorderid INT;
SELECT @minsalesorderid = MIN(SalesOrderID) FROM Sales.
SalesOrderHeader;
BEGIN TRAN;
UPDATE Sales.SalesOrderHeader
SET Freight = Freight * .10
WHERE PurchaseOrderNumber = 'PO18850127500';
GO
-- Rollback transaction if needed
ROLLBACK TRAN;
GO
```

   This script will update the freight for a different specific PurchaseOrderNumber. **Execute the first batch in the query script up to the GO statement. Do not execute the ROLLBACK TRAN statement.** Notice this batch is blocked.

4. Execute the query in **getlocks.sql**. You should see results like in Figure 7-7.

	resource_type	resource_database_id	resource_associated_entity_id	request_mode	request_session_id	request_status	lock_count
1	METADATA	5	0	Sch-S	117	GRANT	1
2	OBJECT	5	1602104748	IS	125	GRANT	1
3	OBJECT	5	1602104748	IX	117	GRANT	1
4	XACT	5	0	X	117	GRANT	1
5	XACT	5	72057594051166208	S	125	WAIT	1

***Figure 7-7.** Blocking due to locks before qualification*

5. Clean up current transactions. Roll back the transactions in **updatefreightpo1.sql** and **updatefreightpo2.sql**. You can do this by executing the **ROLLBACK TRAN** statement in those scripts. Keep all the scripts open.

Example 2: Locking with LAQ

Now let's enable LAQ and see the blocking just disappear with no code changes!

1. Enable READ_COMMITTED_SNAPSHOT with the script **enablercsi.sql.**

2. Run the statements (but not the ROLLBACK TRAN) in the scripts **updatefreightpo1.sql** and **updatefreightpo2.sql.** Notice the second script does not block!

3. Execute the query in **getlocks.sql**. You should see results like in Figure 7-8.

	resource_type	resource_database_id	resource_associated_entity_id	request_mode	request_session_id	request_status	lock_count
1	METADATA	5	0	Sch-S	90	GRANT	1
2	OBJECT	5	1602104748	IX	81	GRANT	1
3	OBJECT	5	1602104748	IX	90	GRANT	1
4	XACT	5	0	X	81	GRANT	1
5	XACT	5	0	X	90	GRANT	1

Figure 7-8. *Locks with Lock After Qualification*

4. You can roll back all transactions and exit all scripts.

Think of this. We have solved two major blocking problems with no code changes. That is the power of optimized locking.

tempdb Resource Governance

Over several releases of SQL Server, we have attempted to provide enhancements toward easing your management of tempdb most notably to reduce or eliminate latch contention on system allocation pages.

One problem we have continued to hear though is about *out-of-control* tempdb growth. While I always believe you should set up tempdb to support autogrow so you don't run out of space in an unexpected situation, it would be great if you could have a more reliable way to control the amount of space used in tempdb by various users.

At the same time, we have had the concept of Resource Governor as far back as SQL Server 2008 to help govern CPU, memory, MAXDOP, and I/O. So why not bring the concept of space usage governance for tempdb using Resource Governor?

That is exactly what we have done in SQL Server 2025. Effectively we have added two new options for a WORKLOAD GROUP for SQL Server 2025: **GROUP_MAX_TEMPDB_DATA_MB** and **GROUP_MAX_TEMPDB_DATA_PERCENT**. You can use this for the default workload group (all users) or a specific workload group you create.

Note Be very careful setting this for the default workload group. You are now setting a restriction on all users. It is like setting a fixed tempdb size with no autogrow for all users.

I asked Dimitri Furman, lead program manager for the SQL Server engine (and one of the most knowledgeable SQL experts I know), the thinking behind this feature. He said, *"Many DBAs had to deal with an out-of-space tempdb at least once in their career. In this SQL Server release, we wanted to build a solution within the database engine itself, so DBAs wouldn't need to spend time and effort on custom watchdog scripts. Extending Resource Governor to govern tempdb space was a natural choice, giving customers the flexibility to implement both simple and advanced configurations. They say that the simplest solution is best, so that's what we did: starting with SQL Server 2025."*

Let's look at an example to see how this works.

Prerequisites for the Example

You can find all scripts in the samples with the book in **the ch7 - Core Engine\tempdbrg** folder. I use SQL Server Management Studio (SSMS) for all script and T-SQL statement execution.

1. You will need to **enable mixed-mode authentication** for the SQL Server instance you are using. This is because the demo will use a new SQL login to demonstrate the tempdb space resource governance feature. You can do this by following the instructions in documentation at `https://learn.microsoft.com/sql/database-engine/configure-windows/change-server-authentication-mode`.

2. In this scenario you want to establish the default size of tempdb to be 512MB across 8 tempdb data files and a 100MB tempdb log file. You can do this by executing the script **settempdbsize.sql** logged in as a sysadmin. This will set the size of tempdb to 512MB across 8 data files and a 100MB log file.

    a. You decide to leave autogrow for tempdb files to avoid any downtime situations, but your goal is to make sure growth does not exceed 512MB because you have carefully planned the usage of tempdb through temporary tables from your developers. In other words, you don't want to *run out of disk space* for the drive holding tempdb.

    b. Execute the script **checktempdbsize.sql** in a SSMS query editor window to verify the size of tempdb. You should see that the size of tempdb is 512MB across 8 data files and a 100MB log file.

3. Load and execute the script **createbigdata.sql** to create a new database called **guyinacubedb** with data logged in as a sysadmin. This will create a database with a larger amount of data for our less experienced users.

Note It is fair I'm poking fun at the `https://www.youtube.com/guyinacube` in this example. But if you don't know, Adam and Patrick are personal friends of mine, so I know they will not be offended. And, Patrick, when you read this, Power BI is *not* the center of the universe!

---

4. Load and execute the script **createuser.sql** logged in a sysadmin. This will create a login and user for our database from the previous step.

5. Load and execute the script **iknowsql.sql** logged in as sysadmin. This will create a database called **iknowsqldb**. This includes a table and stored procedure that will create a temporary table of fixed size that ensures a more controlled use of tempdb.

Examples for tempdb Resource Governance

With everything set up let's walk through examples to see controlled and uncontrolled usage of tempdb. Then let's set up Resource Governor to control tempdb usage and then see it in action.

Example 1: Show controlled tempdb usage

Execute all of these steps logged in as the sysadmin you used to execute scripts in the "Prerequisites for the Example" section.

1. Load and execute the script **processdata.sql**.

   This script will execute a stored procedure in the iknowsqldb database to process data and create a temporary table of fixed size. This script should complete very fast and return a result set of 1,000. If you look at the stored procedure called ProcessData, you will see it creates a temporary table and populates it with a fixed number of rows.

   Let's say you know the maximum concurrent users that will use this procedure and have tested that the size of tempdb you have set, 512MB, is plenty of space to meet the needs of the application.

2. Load and execute the script **checktempdbsize.sql**. This script will check the size of tempdb as you did in the "Prerequisites for the Example" section. Notice the size has not changed, so no growth of tempdb has occurred.

3. Load and execute the script **tempdb_session_usage.sql**. This script uses the following T-SQL statement to show the small amount of tempdb space used by the session that executed the stored procedure for an explicit temporary table:

```
SELECT
    ssu.session_id,
    es.program_name AS appname,
    ssu.user_objects_alloc_page_count * 8 AS user_objects_
    alloc_kb,
    ssu.internal_objects_alloc_page_count * 8 AS internal_objects_
    alloc_kb
```

```
FROM
    sys.dm_db_session_space_usage AS ssu
JOIN
    sys.dm_exec_sessions AS es ON ssu.session_id = es.session_id
WHERE
    (ssu.user_objects_alloc_page_count > 0 OR ssu.internal_
    objects_alloc_page_count > 0);
GO
```

The results should look like Figure 7-9.

	session_id	appname	user_objects_alloc_kb	internal_objects_alloc_kb
1	65	Microsoft SQL Server Management Studio - Query	384	0

Figure 7-9. *Controlled tempdb usage*

You can see for this session that a very small amount of space has been used for "user" objects, which are explicit temporary tables.

Example 2: Show uncontrolled tempdb usage

Now let's look at an example of "uncontrolled" tempdb usage.

For these steps, unless specified, connect with the login **guyinacube**, which was created in the "Prerequisites for the Example" section in the script **createuser.sql** (which has the password you need). To see the examples better, use a tool like SSMS and, in Additional Connection Parameters, set the Application Name to "GuyInCube" like in Figure 7-10.

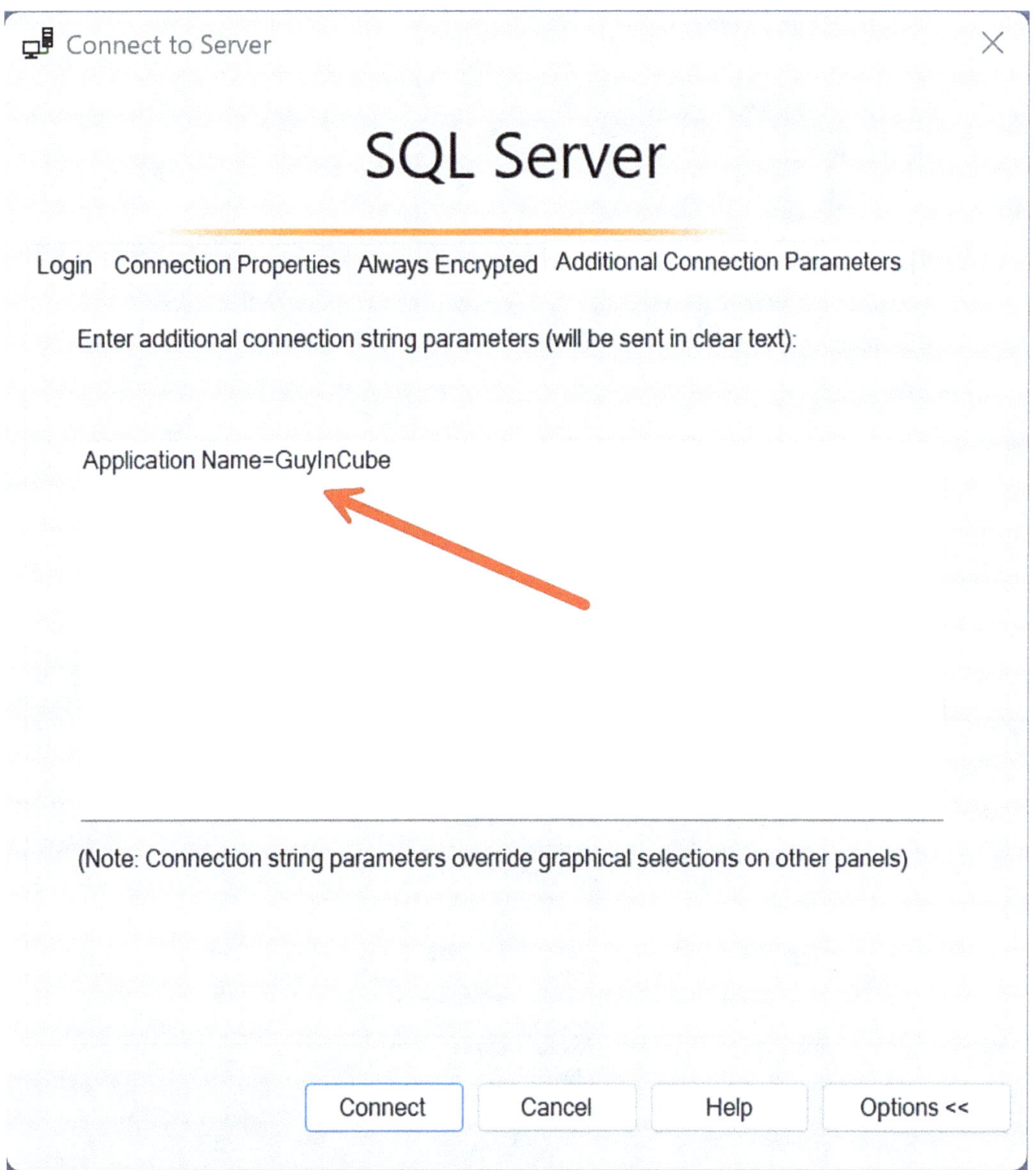

Figure 7-10. *Setting Application Name in SSMS*

1. Load the and execute the script **guyinacubepoorquery.sql** that runs the following T-SQL statement. First, set the option Include Actual Execution Plan in SSMS before executing the query. This script runs the T-SQL statement:

```
USE guyinacubedb;
GO
SELECT * FROM bigtab
ORDER by col2;
GO
```

This script will take several minutes to run and returns 1 million rows. The key result to look at is the execution plan, which looks like Figure 7-11.

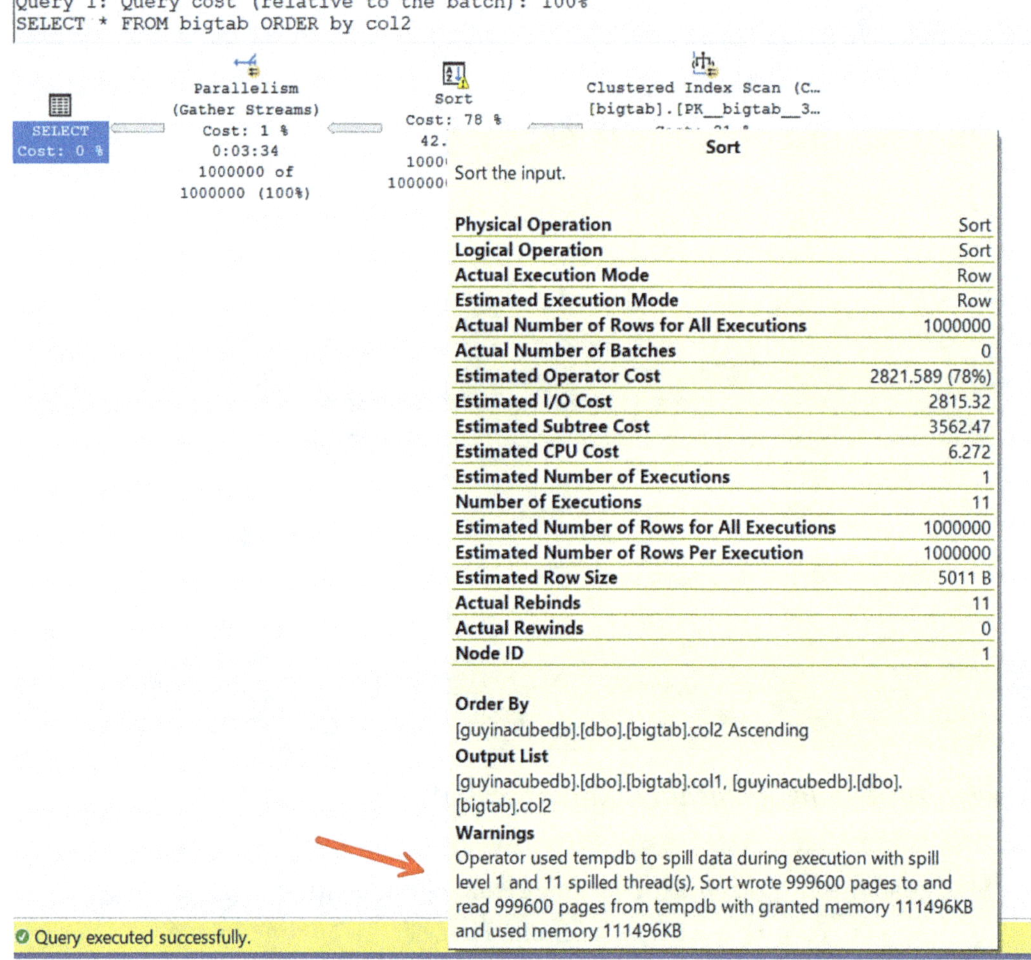

__Figure 7-11.__ Poor query resulting in sort and spill in tempdb

Notice the Sort operator, which results in a spill of tempdb. This is not only a poor query but results in uncontrolled tempdb growth.

2. Load and execute the script **checktempdbsize.sql**. The results should look like Figure 7-12.

	FileName	CurrentSizeMB	FreeSpaceMB	UsedSpaceMB	PhysicalFileName
1	tempdev	1024.000000	1021.187500	2.812500	C:\Program Files\Microsoft SQL Server\MSSQL17.MSS...
2	templog	1000.000000	889.640625	110.359375	C:\Program Files\Microsoft SQL Server\MSSQL17.MSS...
3	temp2	1024.000000	1023.875000	0.125000	C:\Program Files\Microsoft SQL Server\MSSQL17.MSS...
4	temp3	1024.000000	1023.875000	0.125000	C:\Program Files\Microsoft SQL Server\MSSQL17.MSS...
5	temp4	1024.000000	1023.875000	0.125000	C:\Program Files\Microsoft SQL Server\MSSQL17.MSS...
6	temp5	1024.000000	1023.937500	0.062500	C:\Program Files\Microsoft SQL Server\MSSQL17.MSS...
7	temp6	1024.000000	1023.812500	0.187500	C:\Program Files\Microsoft SQL Server\MSSQL17.MSS...
8	temp7	1024.000000	1023.812500	0.187500	C:\Program Files\Microsoft SQL Server\MSSQL17.MSS...
9	temp8	1024.000000	1023.750000	0.250000	C:\Program Files\Microsoft SQL Server\MSSQL17.MSS...

***Figure 7-12.** Uncontrolled tempdb growth*

Notice tempdb has grown to ~8GB, which is not good. This space won't be reclaimed unless you try to shrink tempdb or restart SQL Server.

3. **Connect as the sysadmin login** and run the script **tempdb_session_usage.sql**.

Your results should look similar to Figure 7-13.

	session_id	appname	user_objects_alloc_kb	internal_objects_alloc_kb
1	54	SQLServerCEIP	0	448
2	65	Microsoft SQL Server Management Studio - Query	384	0
3	73	GuyInCube	0	8001024

***Figure 7-13.** Uncontrolled tempdb usage*

You can see the GuyInCube application has chewed up ~8GB of "internal" objects, which means tempdb usage based on things like spills for sort operations. At this point, an administrator does not know what the query is doing, just that it is chewing up tempdb space out of control. You could start a conversation with developers of this application, but now you also have a different way to control the application's tempdb usage.

Example 3: Set up Resource Governor to control tempdb space usage

Connect as the same sysadmin login you used to set up this example for all the steps in this example.

1. Let's reset tempdb by executing the script **settempdbsize.sql** and restart SQL Server.

2. Load and execute the script **setuprg.sql** that executes the following T-SQL statement:

```
-- Enable Resource Governor if not already enabled
ALTER RESOURCE GOVERNOR RECONFIGURE;
GO

-- Create a new Workload Group with only GROUP_MAX_TEMPDB_DATA_
MB option
CREATE WORKLOAD GROUP GroupforUsersWhoDontKnowSQL
WITH (GROUP_MAX_TEMPDB_DATA_MB = 100);
GO

-- Apply the changes
ALTER RESOURCE GOVERNOR RECONFIGURE;
GO
```

   This script will set up a new workload group and restrict tempdb usage for this group to 100MB.

3. Load and execute the script **classifierfunction.sql.** This script runs the following T-SQL statements:

```
USE master;
GO
-- Create a classifier function
CREATE FUNCTION dbo.ResourceGovernorClassifier()
RETURNS SYSNAME
WITH SCHEMABINDING
AS
BEGIN
    DECLARE @WorkloadGroup SYSNAME;
```

```
    -- Example logic: Assign sessions based on application name
    IF APP_NAME() = 'GuyInCube'
    BEGIN
        SET @WorkloadGroup = 'GroupforUsersWhoDontKnowSQL';  --
Assign to your custom workload group
    END
    ELSE
    BEGIN
        SET @WorkloadGroup = 'default';  -- Assign to the default
workload group
    END
    RETURN @WorkloadGroup;
END;
GO
-- Register the classifier function with Resource Governor
ALTER RESOURCE GOVERNOR WITH (CLASSIFIER_FUNCTION = dbo.
ResourceGovernorClassifier);
GO
-- Apply the changes
ALTER RESOURCE GOVERNOR RECONFIGURE;
GO
```

This script will create a new classifier function so that when the application GuyInCube logs in, it will be assigned to the workload group to restrict tempdb usage.

We are now set up to show how tempdb can be controlled even for users that execute queries that normally would result in uncontrolled growth.

Example 4: Test tempdb is now limited for the workload group

For these steps, unless specified, connect with the login **guyinacube. You must first disconnect the user guyinacube** and reconnect, so the login is placed in the correct workload group.

1. As before, load and execute the script **guyinacubepoorquery.sql**.

   The query will fail immediately with the following error:

   ```
   Msg 1138, Level 17, State 1, Line 3
   Could not allocate a new page for database 'tempdb'
   because that would exceed the limit set for workload group
   'GroupforUsersWhoDontKnowSQL', group_id 256.
   ```

   Now the developers for this application will need to work to tune their query to ensure it does not use uncontrolled tempdb.

Note You can see here this is a different error than you might normally see when you "run out of space" in tempdb, which is typically Msg 1105 (scary that I knew that by memory).

---

2. Load and execute the script **checktempdbsize.sql**. Notice that the size of tempdb has not grown beyond the 512MB limit set as before.

3. **Connect as the sysadmin login** you have been using. Load and execute the script **checktempdbrg.sql**. This script executes the following T-SQL statement:

   ```
   SELECT name,
          tempdb_data_space_kb, peak_tempdb_data_space_kb,
          total_tempdb_data_limit_violation_count
   FROM sys.dm_resource_governor_workload_groups;
   GO
   ```

   The results should look similar to Figure 7-14.

	name	tempdb_data_space_kb	peak_tempdb_data_space_kb	total_tempdb_data_limit_violation_count
1	internal	216	216	0
2	default	0	0	0
3	GroupforUsersWhoDontKnowSQL	72	102408	2

***Figure 7-14.** tempdb resource governance*

You can see from these results the workload group has been limited to a peak of 100MB.

tempdb resource governance is a great example of a feature designed to improve application availability with no code changes required.

Other tempdb Enhancements

We have also built other enhancements for tempdb to improve concurrency and performance.

ADR in tempdb

As I wrote earlier in this chapter on optimized locking (OL), Accelerated Database Recovery (ADR) supports versions inside the database. Independent of OL, ADR provides shorter recovery, quick rollback, and reduced log space usage.

tempdb is rebuilt each time SQL Server is restarted, so fast recovery with ADR is not a benefit. But transactions are still supported and logged for tempdb, so fast rollback and reduced log space usage are of benefit.

Therefore, we now support enabling ADR for tempdb for SQL Server 2025. Simply execute the following T-SQL statement

```
ALTER DATABASE tempdb SET ACCELERATED_DATABASE_RECOVERY = ON;
GO
```

and restart SQL Server. One big benefit of ADR is reduced log space, so using ADR can help you reduce log space growth in tempdb.

Any use of ADR should be well tested including any transaction performance or space usage ramifications for versions in the database, in this case inside tempdb.

tmpfs Support for tempdb on Linux

In a galaxy far, far away, SQL Server supported the tempdb database to be in RAM. This feature was removed and has never come back. I've had customers over the years ask me to bring back this feature. I've never quite seen the real benefit for this given

- tempdb doesn't use the normal checkpoint algorithms as user databases (it can occur) because tempdb is recreated at restart.
- tempdb uses delayed durability by default for transaction commits.

So the only scenarios where I/O might be an issue for tempdb are as follows:

- Memory pressure occurs and any active database page for a temporary table must be flushed and read from disk.
- Operations like sorts result in spills for tempdb. But these should be avoided anyway.

So there may be a case for bringing back tempdb in RAM, but not in our plans right now. The implementation (way back in SQL Server 4.X days) was a special implementation.

However, turns out Linux supports a file system called **tmpfs**, which is backed by virtual memory (`https://www.kernel.org/doc/html/latest/filesystems/tmpfs.html`).

With SQL Server 2025, we now support the use of tempdb files (data and log) on tmpfs file systems for Linux. The max size of tmpfs is the size of virtual memory for the OS (RAM + swap), but you can control its maximum size.

SQL Server 2025 also lets you control the maximum size of tempdb usage for tmpfs through setting maximum sizes of tempdb files or the maximum size using the Linux **mount** command.

You can read the documentation for how to configure and set up tempdb to use tmpfs for Linux and containers at `https://learn.microsoft.com/sql/linux/sql-server-linux-tmpfs-tempdb`.

Note I would personally be careful using this feature and ensure it will really benefit your workload. This is because RAM may be a precious resource for your computer or virtual machine. If you do use this feature, consider sizing the use of tempdb for tmpfs to avoid the use of swap space since this results in I/O and could defeat the purpose of using this feature. In addition, remember that if you use this feature, you can use up RAM for *both* buffer pool pages on disk and in memory so you may have less memory than you are expecting for SQL Server buffer pool management.

Query Optimization and Execution

There are several enhancements that benefit query performance either through query optimization or execution methods. There are a range of features here that can benefit different types of workloads.

Optimized sp_execute_sql

The system procedure **sp_execute_sql** was introduced in SQL Server 7.0 to aid developers for *parameterization* of ad hoc queries. If you have queries that "look the same" except for say values in the WHERE clause, it would be better in many cases to have one execution plan for these combinations vs. filling up the plan cache. This also reduces query compilations, which can chew up CPU usage. In SQL Server 2000, we also introduced the concept of *auto-parameterization*, so developers did not always need to specify in their code to parameterize the query.

Still, I believe a best practice is to use a stored procedure or parameterize often executed queries. This technique doesn't just benefit performance but can also prevent problems like SQL injection attacks. In fact our providers will use sp_execute_sql if you use APIs to parametrize an ad hoc query.

Note I totally understand the issue of parameter-sensitive plans. We created a feature to assist in this issue in SQL Server 2022 called Parameter Sensitive Plan Optimization (PSPO), which we have enhanced in SQL Server 2025, which you will learn later in this chapter. You can learn more at `https://learn.microsoft.com/sql/relational-databases/performance/parameter-sensitive-plan-optimization`.

Here is a code example using C#, which will result in the provider using sp_execute_sql:

```
using System;
using System.Data.SqlClient;
class Program
{
```

```
    static void Main()
    {
        string connectionString = "Server=your_server;Database=
        your_db;Integrated Security=true;";
        string query = "SELECT * FROM dbo.YourTable WHERE Id = @Id";
        using (SqlConnection connection = new SqlConnection
        (connectionString))
        using (SqlCommand command = new SqlCommand(query, connection))
        {
            command.Parameters.AddWithValue("@Id", 42);
            connection.Open();
            using (SqlDataReader reader = command.ExecuteReader())
            {
                while (reader.Read())
                {
                    Console.WriteLine(reader["Name"]);
                }
            }
        }
    }
}
```

For this code, the provider sends this T-SQL to the server:

```
exec sp_executesql
    N'SELECT * FROM dbo.YourTable WHERE Id = @Id',
    N'@Id int',
    @Id = 42;
```

This is a really good story for developers. Here is the problem. Unlike the execution of stored procedures, a compile lock (which looks like an OBJECT lock on the stored procedure) is not acquired for sp_execute_sql. Therefore, for large concurrent executions of sp_execute_sql, you can defeat the purposes of a single cached plan because multiple cached plans can be created.

In SQL Server 2025 (and with Azure SQL), you can use the new database option OPTIMIZED_SP_EXECUTESQL set with ALTER DATABASE to enforce a compile lock on sp_execute_sql specific to the query being passed in with this system procedure.

It is possible you have not seen this as an issue. But if you use parameterized queries heavily without stored procedures, this might be an issue for an unnecessary increase in plan cache for which this feature can help.

The Dynamic Management View (DMV) **dm_exec_cached_plans** can be used to see if you are having this issue. I've built an example for you to see what this looks like at https://github.com/microsoft/bobsql/tree/master/demos/sqlserver2025/engine/performance/Optimized%20sp_executesql.

Cardinality Estimation (CE) Feedback for Expressions

In SQL Server 2014 we made a difficult decision—but we felt the right decision—to update how cardinality estimation works for SQL Server in query optimization. This change correlates with dbcompat 120. We call this the **new CE model**. Any use of this dbcompat level will result in a different algorithm for cardinality estimation, one of the most important aspects for query performance.

I covered all of these details at length in Chapter 5 of the book *SQL Server 2022 Revealed* as part of the discussion of an enhancement called **Cardinality Estimation (CE) Feedback**. In the book I called out two important resources to get a deep dive on this topic:

- A blog we wrote on why we built a new CE model at https://cloudblogs.microsoft.com/sqlserver/2014/03/17/the-new-and-improved-cardinality-estimator-in-sql-server-2014/
- An incredible detailed whitepaper written by the famous Joe Sack at https://download.microsoft.com/download/d/2/0/d20e1c5f-72ea-4505-9f26-fef9550efd44/optimizing%20your%20query%20plans%20with%20the%20sql%20server%202014%20cardinality%20estimator.docx

CE Feedback in SQL Server 2022 was the capability of SQL Server to "learn" from feedback of query execution using the query store to improve query performance for queries that are sensitive to the **new CE model**. Scenarios covered by CE Feedback include correlation, join containment, and a concept called row goals. You can read about how CE Feedback works for these scenarios and examples for them at https://learn.microsoft.com/sql/relational-databases/performance/intelligent-query-processing-cardinality-estimation-feedback. We have found working with customers that some scenarios are not covered by CE Feedback, namely, a concept called *expressions*. An example could be a query where different join containment choices affect different parts of a query (e.g., two different joins).

To add coverage for more scenarios, we have added a new database scoped configuration option for CE Feedback called CE_FEEDBACK_FOR_EXPRESSIONS. Like CE_FEEDBACK, this option requires dbcompat 160 or higher (but is not available in SQL Server 2022). You can read more details on exact scenarios, diagnostics, and examples for CE Feedback for expressions including the caching of "expression" by the engine at `https://learn.microsoft.com/sql/relational-databases/performance/intelligent-query-processing-ce-feedback-for-expressions`.

Tip Just like CE_FEEDBACK, CE Feedback for expressions will not work if you are using the database scoped configuration option LEGACY_CARDINALITY_ESTIMATION. And I realize this option is very important as it allows you to use the latest dbcompat levels. If you rely on this in production, my recommendation is to get a test workload and server to try to turn this option off and turn on both CE_FEEDBACK and CE_FEEDBACK_FOR_EXPRESSIONS. Use the query store to compare differences for queries with and without these options.

Optional Parameter Plan Optimization (OPPO)

CE Feedback is part of a family of features we call intelligent query processing (IQP) (see `https://aka.ms/iqp`). Another feature that is part of IQP I mentioned earlier in this section is called Parameter Sensitive Plan Optimization (PSPO). PSPO provides a mechanism to provide the optimal execution plan if multiple parameters could affect different plans. PSPO is enabled by default with dbcompat 160 or higher.

There is one scenario we have seen that PSPO doesn't handle with what we call *optional* parameters. PSPO can work with different parameter value ranges, but what if the value can be NULL (hence optional)?

Consider a statement like the following:

```
SELECT column1,
       column2
FROM Table1
WHERE column1 = @p
      OR @p IS NULL;
```

In this case, PSPO can't decide which "range" to use for the plan so has to use a table scan even if there is a valid index on Table1.column1.

Now we have extended PSPO, called Optional Parameter Sensitive Plan Optimization (OPSPO), to account for this example. With OPSPO, we can choose a table scan if the parameter value is NULL and use the index or a better plan if the value is NOT NULL.

OPSPO only requires dbcompat 170 but can be turned on and off with the database scoped configuration value OPTIONAL_PARAMETER_OPTIMIZATION. Learn more at `https://learn.microsoft.com/sql/relational-databases/performance/optional-parameter-optimization`.

DOP Feedback

Another IQP feature we introduced in SQL Server 2022 was Degree of Parallelism (DOP) feedback. Similar to CE Feedback, the query optimizer in partnership with the query store can evaluate query execution and make adjustments to DOP. The goal of DOP Feedback is the *optimal* DOP to use for a query instead of using MAXDOP. The result can be less CPU resources to execute a query but with the same overall duration. In SQL Server 2022, DOP Feedback required dbcompat 160 *and* the database scoped configuration option DOP_FEEDBACK to be on, which was not on by default.

Now in SQL Server 2025, DOP_FEEDBACK is ON by default with dbcompat 160 or higher.

Read more to see if DOP Feedback can help your query workload at `https://aka.ms/dopfeedback`.

Query Management

There are a few enhancements for SQL Server 2025 to manage query performance including columnstore indexes, the query store, and persisted statistics.

Columnstore Index Improvements

Columnstore indexing is such an amazing feature built into SQL Server, yet I still see it underused. It is not for every workload, but if you are using SQL Server for any type of "analytic" set of queries involving querying a large amount of data, give columnstore indexes a spin. It already comes with the product and requires no code or query changes.

In SQL Server 2025, we have added a few enhancements to help you manage columnstore indexes to accelerate performance and provide maximum availability:

- **Non-clustered columnstore indexes can be ordered.**

  When you create a columnstore index, creating the index using ordering is a critical part to achieve maximum performance. According to our documentation at `https://learn.microsoft.com/sql/relational-databases/indexes/ordered-columnstore-indexes`, an ordered columnstore index avoids segment overlaps ensuring less segments are required for queries using the index. We started support for ordered clustered columnstore indexes in SQL Server 2022 and additionally now have support in SQL Server 2025 for ordered non-clustered columnstore indexes.

- **Ordered columnstore indexes can be created or rebuilt online.**

  Even with the support for ordered columnstore indexes, you were previously not able to create or rebuild the indexes online. SQL Server 2025 supports online columnstore index creation or rebuild. It is important to know that in order to achieve this, we have to use tempdb storage and you must specify MAXDOP = 1.

- **Improved shrink operations.**

  SQL Server 2025 now allows shrink operations for data pages that include columns of types varchar(max), nvarchar(max), and varbinary(max) that are part of a columnstore index.

Query Store on Secondary Replicas

I remember the excitement when I found out in the planning phases of SQL Server 2022 (Project Dallas) that we would release the ability to use a query store on secondary replicas, specifically to track the performance of queries on readable secondary replicas.

By now you may know that when we released SQL Server 2022 GA, this feature was only available under a trace flag, 12606, indicating it was not meant for production. In all honesty, we thought shortly after we released the product we could lift the trace flag in a Cumulative update and all would be right with the world. That didn't happen to the disappointment of us and most notably you, our customers.

Unfortunately, for SQL Server 2022, that trace flag remains to this day. The good news is that the trace flag is gone for SQL Server 2025, and the feature is fully supported in production.

I remember asking our engineering team how they were going to pull off this feature when secondary replica databases by their nature can't be modified by users. I went back from my SQL Server 2022 content and found Figure 7-15.

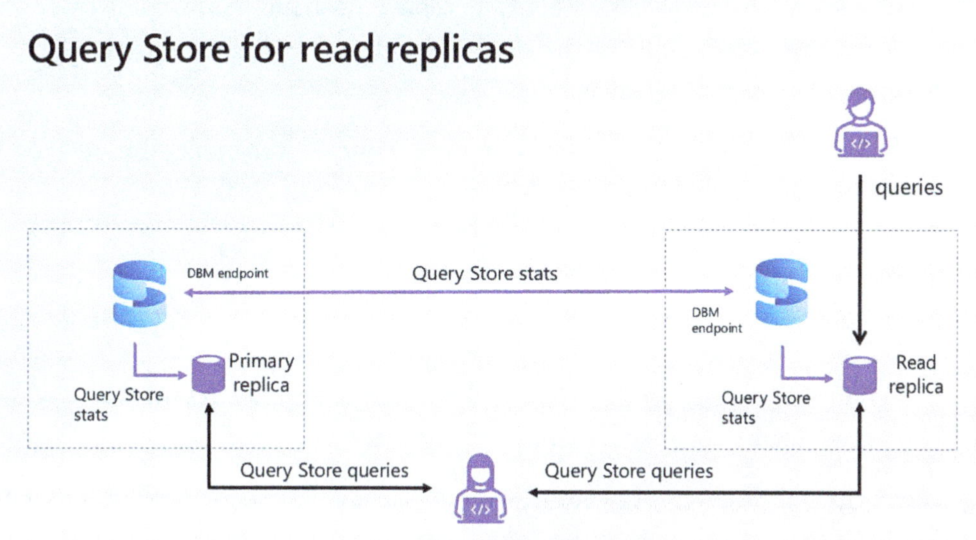

Figure 7-15. *Query store on read replicas*

Here is how we pull off this magic. The query store is a series of memory structures and system tables. For a primary replica, we just treat this like a normal query store. For the secondary replica, when changes to the query store are made in memory structures, we **stream** these changes *back to the primary* on the same channel used for AG communications.

The primary recognizes these streams and writes the information into the query store system tables on the primary database. The query store system tables have been enhanced to support a **replica ID**, so you know which query was executed on which replica. Since modifications to system tables on the primary replica are logged, they are sent like other transactions for the AG to the secondary where they are hardened and redone. Now anyone connecting to *any replica* in the AG can see the query store across all replicas!

This option is "all or nothing." You enable the query store on secondary replicas with one command on the primary, and it affects **all** replicas.

Here are a few notes about this feature:

- The results may have some lag as we have to stream query store information back to the primary and we don't do this immediately so as to not flood the communication channel between replicas.
- Query store information sent back to replicas also has a lower priority than normal "AG traffic" so we don't disrupt any operations with the performance of the replicas.
- All of the normal considerations for tuning query store configuration information on the primary still exist.
- Cleanup happens on the primary query store, and since these changes are logged, they will be replicated to each secondary query store.
- When a failover occurs, SQL Server knows which replica is now the primary to take over query store write operations.

Finally, Erin Stellato is smiling now that this feature is fully supported for production workloads.

Persisted Stats on Readable Secondaries

The database option AUTO_CREATE_STATISTICS was specifically designed to help query performance when statistics didn't exist (say through an index) and the creation of a separate statistic would help cardinality estimation and therefore query performance. With this in mind, if you are running read queries on secondary replicas, they are in many cases different queries that you run on the primary replica (a.k.a. offload read workloads). The problem is that since the secondary replica is "read-only," we cannot create statistics in the secondary replica database. Therefore, in SQL Server 2014 we added the concept of temporary statistics on secondary replicas. Statistics on these replicas are created in tempdb with special names so the optimizer on secondary replicas can recognize these and use them.

Here is the problem. What if the secondary gets restarted? tempdb gets recreated, and all the temporary statistics are lost. You effectively have to "start over." So, in SQL Server 2025, we have introduced the concept of *persisted* statistics for secondary replicas.

If you enable the query store for secondary replicas, you will have now enabled the ability for secondary replicas to "talk" back to the primary. Now if a temporary statistic is created, the secondary will send this information back to the primary, which will persist this statistic into system tables in the primary database. That is a logged change, so it will make its way back to the secondary. Queries on the secondary will continue to use the temporary statistics until the query needs to be recompiled or the secondary is restarted. This also applies to any statistics that are automatically created and need to be updated. Any update would go back to the primary and make its way back to the secondary.

You will need to enable the query store on your secondary for this to take effect even if you don't plan to use the query store as this enables the communication path back to the primary. You can learn more at `https://learn.microsoft.com/sql/relational-databases/performance/persisted-stats-secondary-replicas`.

Availability

As I mentioned earlier in the chapter, this section on availability covers both High Availability (HA) and Disaster Recovery (DR). The first part of this section covers enhancements for our flagship HADR feature, Always On Availability Groups. The second part of this section covers DR features centered around enhancements for backups. As a reference point, all new availability features are documented at `https://learn.microsoft.com/sql/sql-server/what-s-new-in-sql-server-2025#availability`.

Always On Availability Groups (AGs)

Sorry, Allan Hirt, but I'm going to use AGs in this chapter because I find it silly to type out the full name every time. AG enhancements in SQL Server 2025 are one of the *hero stories* for the release. I remember vividly looking at our Azure DevOps (ADO) repo for our improvements in the engine coming to SQL Server 2025. I had to do a double-take as I saw well over six to seven improvements for AGs. I have to admit I was like "Where did this all come from?" As I looked at all of these enhancements, I caught the name Dong Cao. Dong is one of the most talented principal software engineers for the SQL Server engine and has become the "de facto" expert on our AG code.

I emailed Dong asking him about these enhancements. He told me he and his team had been collecting various enhancements since SQL Server 2022 based on direct customer feedback. There was nothing "major" on his list, but altogether it represents an impressive collection of enhancements. And the enhancements were not based on a theory. They were based on actual evidence of customer problems and requests.

The collection of enhancements covers everything from improved failover, configuration, tuning, diagnostics, and Distributed Availability Group (DAG) enhancements.

Creating examples for these enhancements is very difficult. So based on a pointer from my friend Sean Gallardy in engineering, I connected with James Ferebee. James is a senior escalation engineer in our CSS organization (which is fun for me as that is my roots). What an incredible knowledge James has for SQL Server and AGs! We worked through many of these AG enhancements on devising examples to showcase how they work and when they would be useful for customers. Unfortunately, many of these are just not possible to give you a cookbook to show a step-by-step example, but in some cases, we can show you how we observed the improvement.

If you are going to explore any of these yourself, you need an AG and a Windows Server Failover Cluster (WSFC) as well. For me, when I need to do this, I use Azure Virtual Machines. Here are some resources I used to implement a WSFC with one domain controller, two SQL Server 2025 instances (a primary and a secondary), and a cloud witness. If you go down this path, it takes time, but when you get it running, it is well worth it:

`https://learn.microsoft.com/azure/azure-sql/virtual-machines/windows/availability-group-manually-configure-prerequisites-tutorial-multi-subnet?view=azuresql`

`https://learn.microsoft.com/azure/azure-sql/virtual-machines/windows/availability-group-manually-configure-tutorial-multi-subnet?view=azuresql`

Note When SQL Server 2025 becomes GA, use the new portal experience. It will save you tons of time: `https://learn.microsoft.com/azure/azure-sql/virtual-machines/windows/availability-group-azure-portal-configure`.

Failover Improvements

Here is the good news about failover improvements for SQL Server 2025. The only thing you need to do is upgrade! All the improvements are just built into the product. No SQL Server configuration required. Let's talk about each of these and why they improve your experience with AGs in SQL Server 2025.

Fast failover for persistent health issues

SQL Server in coordination with WSFC has a sophisticated method to determine whether to fail over to a secondary node. The method of detection is the same for a Failover Cluster Instance or an Always On Availability Group. Without going into an incredible depth of how this works, the basic methods are a **health check timeout** and **lease timeout**. The health check timeout uses a SQL internal stored procedure to check the health of SQL Server. The lease timeout is a special shared memory communication between the SQL Server engine and a local process called the cluster resource. (All the details are well documented at `https://learn.microsoft.com/sql/database-engine/availability-groups/windows/availability-group-lease-healthcheck-timeout`.)

The default behavior for WSFC is if one of these timeouts is triggered, the cluster will first attempt to **restart** the *resource* one time, which in this case is the primary replica of the AG (not the instance, but the AG resource). If the restart allows health and lease checks to perform successfully, no further action is required. However, if the restart does not, a failover will be triggered.

Note According to Sean Gallardy, who is also one of the best experts I know on AGs (and a friend despite being a Philadelphia Eagles fan), *"Restarting the AG resource in WSFC will only cause the AG resource in the cluster (backed by hadrres.dll) to redo the startup sequence for that specific AG on the cluster side (the hadrres.dll implemented monitor). It will not restart SQL Server. For an FCI, it will restart the instance."*

For a *transient* type of problem (maybe SQL Server had some odd responsive issue that a restart of the AG resource resolves), this is a good mechanism as it avoids unnecessary failover.

However, what if the problem is persistent and a restart of the AG resource won't solve the problem? How can you as a user know? One way to be sure that a failover simply happens every time to avoid an unnecessary delay is to change the **RestartThreshold** value for the cluster resource (`https://learn.microsoft.com/previous-versions/windows/desktop/mscs/resources-restartthreshold`). This setting has been around a long time for clusters, and we have supported this behavior by just letting the cluster handle the use and execution of this setting. We found some scenarios when a user would set this value to 0 and our resources had problems honoring a "fast failover" correctly. Then we also discovered a few other issues even when the value was 1. Therefore, we enhanced SQL Server to better integrate in our code for the settings of RestartThreshold.

I asked James what was the origin of this problem. His research indicated that we would have customers with a persistent I/O problem, which caused our health check timeout to fail but a restart didn't help. The question now is, should you change it?

My recommendation is to use all the tools at your disposal including cluster logs, ERRORLOG files, and Always On Health Extended Events to see a history of whether a restart has occurred and solved unnecessary failovers. James and I devised a way to trigger this condition by using the famous sysinternals tool, Process Explorer (`https://learn.microsoft.com/sysinternals/downloads/process-explorer`), to suspend the SQLSERVR.EXE process. This triggered a lease timeout (this triggers before the health check) and an attempt to restart the AG resource. But since we suspended the SQL Server process, the restart of the AG resource didn't work.

What we observed in the cluster logs was something like this

```
WARN  [RCM] Queueing immediate delay restart of resource adwag in 500 ms.
```

and then 30 seconds later

```
INFO  [RCM] Resource adwag is causing group adwag to failover
```

Then, we got everything back healthy again and followed the instructions in our documentation at `https://learn.microsoft.com/en-us/sql/database-engine/availability-groups/windows/failover-and-failover-modes-always-on-availability-groups#fast-failover-for-persistent-health-issues` to set the RestartThreshold to 0.

We tried the same scenario again and now saw this behavior in the cluster logs:

```
ERR [RES] SQL Server Availability Group <adwag>: [hadrag] Lease renewal
failed with timeout error
ERR [RES] SQL Server Availability Group <adwag>: [hadrag] The lease is expired.
ERR [RES] SQL Server Availability Group <adwag>: [hadrag] Availability
Group lease is no longer valid
```

```
WARN [RHS] Resource adwag IsAlive has indicated failure.
INFO [RCM] Resource adwag is causing group adwag to failover.
```

All of this happened immediately, so the failover was instant.

My recommendation is that if you see restart messages like above with no subsequent failovers, then you probably shouldn't change the **RestartThreshold** value. However, you still need to investigate why the transient restarts are happening. If you see failovers, then I would change the RestartThreshold to 0. Either way, you should investigate why you are seeing health check or lease timeouts. Use this documentation as a starting point, `https://learn.microsoft.com/troubleshoot/sql/database-engine/availability-groups/troubleshooting-availability-group-failover`.

Note You will see below later in this section we have added in SQL Server 2025 more health check timeout diagnostics into the cluster log.

The mystery of the Not Synchronizing state

The WSFC and AG system is a giant, complicated *state machine*. To ensure the AG system that is set up for auto-failover knows who the "boss" is or primary, we need *quorum*. If quorum is lost, the cluster must go offline to avoid a split-brain scenario. Read more on this concept at `https://learn.microsoft.com/en-us/sql/sql-server/failover-clusters/windows/wsfc-quorum-modes-and-voting-configuration-sql-server#VotingandNonVotingNodes`.

In my setup, I have two nodes and a cloud witness for the cluster. If two of the three resources are offline or not reachable, you can lose quorum, which will cause the cluster to go offline. If the cluster is offline, the AG is taken offline, which has a major impact to your application. If the problem is permanent, you will need to take action. If the problem is *transient*, the cluster can come back online on its own.

Let's take an example. In my setup, I have two SQL Server replicas, a primary and a secondary, and a cloud witness. I have a single AG group with one database, based on AdventureWorks, in the AG group. To simulate a loss of quorum, I did the following:

- Set up a firewall rule on my primary node VM inside Windows to block outbound network traffic on port 443. This is the port used to "talk" to the cloud witness in Azure.
- Shut down my secondary node VM in Azure.

Now two out of the three resources in the cluster are not available, so a quorum loss occurs. What does this look like?

- You can access the primary SQL Server instance but not the database. If you attempt to access the database, you will get an error like this:

```
Msg 983, Level 14, State 1, Line 1
Unable to access availability database 'AdventureWorks' because
the database replica is not in the PRIMARY or SECONDARY role.
Connections to an availability database is permitted only when
the database replica is in the PRIMARY or SECONDARY role. Try the
operation again later.
```

- The database, AG health, AG role, and status have specific characteristics.

  Figure 7-16 shows a great view of what the database and AG look like in SSMS.

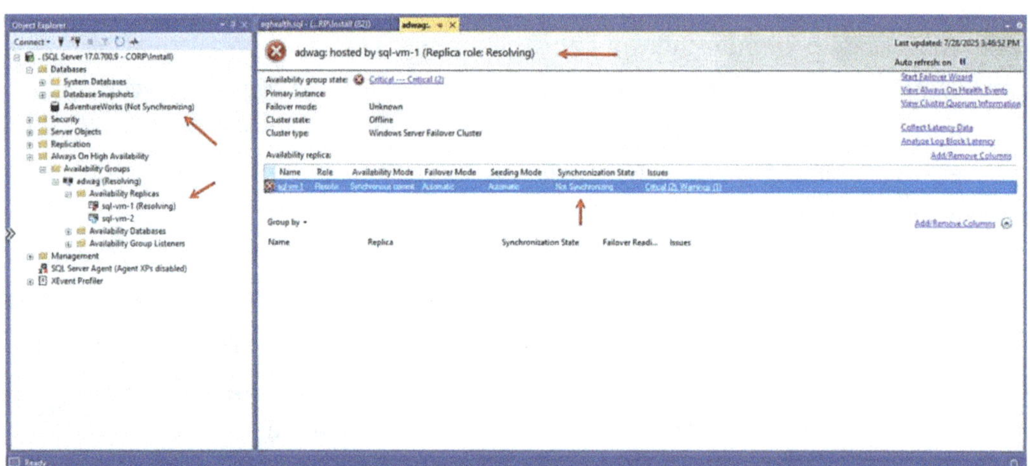

Figure 7-16. *A view of AG state when quorum loss occurs*

- I can also see the state of the database and AG through a query like this:

```
SELECT ag.name AS [AG Name],
ags.primary_replica AS [Primary Replica],
ags.synchronization_health_desc AS [AG Health],
```

```
ar.replica_server_name AS [Replica Name],
ar.availability_mode_desc AS [Availability Mode],
ar.failover_mode_desc AS [Failover Mode],
ars.role_desc AS [Replica Role], d.name AS [Database Name],
drs.synchronization_state_desc AS [Sync State],
drs.is_suspended AS [Is Suspended],
drs.suspend_reason_desc AS [Suspend Reason] FROM
sys.availability_groups ag JOIN
sys.dm_hadr_availability_group_states ags
ON ag.group_id = ags.group_id
JOIN sys.availability_replicas ar
ON ag.group_id = ar.group_id JOIN sys.dm_hadr_availability_
replica_states ars
ON ar.replica_id = ars.replica_id AND ars.group_id = ag.group_id
JOIN sys.dm_hadr_database_replica_states drs
ON ar.replica_id = drs.replica_id AND ag.group_id = drs.group_id
JOIN sys.databases d ON drs.database_id = d.database_id
ORDER BY ag.name, ar.replica_server_name, d.name;
```

The results look like 7-17.

	AG Name	Primary Replica	AG Health	Replica Name	Availability Mode	Failover Mode	Replica Role	Database Name	Sync State	Is Suspended	Suspend Reason
1	adwag	NULL	NOT_HEALTHY	sql-vm-1	SYNCHRONOUS_COMMIT	AUTOMATIC	RESOLVING	AdventureWorks	NOT SYNCHRONIZING	0	NULL

Figure 7-17. *T-SQL results when quorum loss occurs*

- From a cluster perspective, you can see the cluster is down in the Failover Cluster Manager like in Figure 7-18.

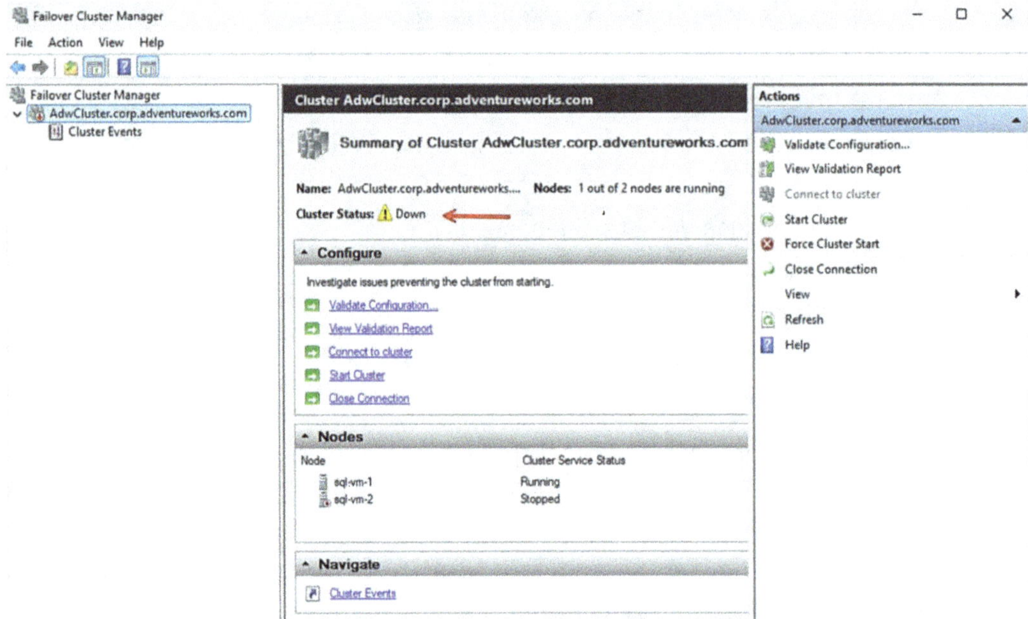

__Figure 7-18.__ Cluster state in the Failover Cluster Manager when quorum loss occurs

- In addition, there are specific cluster events that show the quorum loss problem like in Figure 7-19.

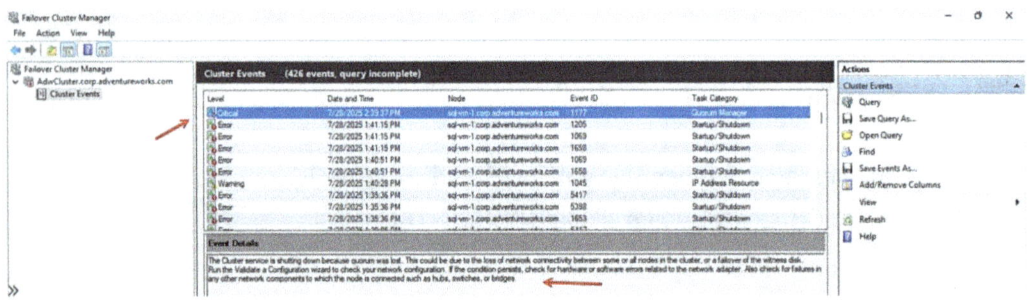

__Figure 7-19.__ Cluster events when quorum loss occurs

Once I brought the second node online and removed the firewall rule to prevent blocking the cloud witness, the cluster came back online, and everything was healthy.

Now that you know what a quorum loss can look like, what does this have to do with the *mystery* of Not Synchronizing? Several customers, including the famous MVP Ola Hallengren, reported to us that a quorum loss would occur for a short period of time (for various reasons), but the cluster did not come back online, and the AG remained offline with the status of Not Synchronizing. One of the interesting aspects of this problem is that customers would be able to just restart the SQL Server instance of the node that was online (primary or secondary) and the cluster would then come back online. That was not expected.

After several iterations of working with customers and studying all the cluster logs and details, our team found an interesting scenario where our code would not react properly when the quorum loss was in this transient state.

This led to the enhancement built into our code to react to quorum loss scenarios like this to ensure the AG would come back online that could result in a failover. But the AG would come back online without manual intervention.

Note We have backported this enhancement in CUs for SQL Server 2019 and 2022. Learn more at `https://learn.microsoft.com/en-us/troubleshoot/sql/releases/sqlserver-2019/cumulativeupdate30#3560254` and `https://learn.microsoft.com/en-us/troubleshoot/sql/releases/sqlserver-2022/cumulativeupdate17#3560259`.

I showed you how I created a quorum loss, but the timing of this issue is so delicate it is not simple for you to manually recreate this. According to James our team had to use a series of very manual precise steps to recreate the problem, which included internal "failpoint" logic in debug versions of our code. I told you, the AG and WSFC system is a complicated state machine.

Asynchronous page request dispatching improvement

Apart of the AG state machine, we have to ensure that replicas are "in sync" when synchronous operations need to occur after a failover event. This concept is called finding a *common recovery point* to synchronize to. The common recovery point is required to allow the AG to be in a synchronized state. As part of this operation, it is possible a replica will need to *revert* a transaction that has already been redone. This revert is also known as *undo of redo*. When this process occurs, it is possible the log records don't exist anymore on the replica because transaction(s) were already redone. So, to perform the revert, pages must be dispatched from the other replica (the new secondary). Prior to SQL Server 2025, the collection of these pages was done completely in a serial fashion. We discovered that in some scenarios where there was a network latency between replicas, the serial processing of the pages (especially if there was a lot of pages to process) would slow down getting back to the common recovery point.

In SQL Server 2025, we can perform these operations asynchronously and in parallel. We first go ask the other replica for all pages needed in one request. Then we start processing pages as they are received. Processing of each page is done in serial, but our request to get all pages is done asynchronously. While this behavior is built-in, it may not improve any performance if the latency between replicas is low. While we don't believe having this on by default can cause any harm, there is a way to disable this. Read more at `https://learn.microsoft.com/sql/database-engine/availability-groups/windows/failover-and-failover-modes-always-on-availability-groups#asynchronous-page-request-dispatching-improvement`.

Tuning, Configuration, and Diagnostics

There are several enhancements designed to allow you to tune, configure, and enhance diagnostics.

Control communication flow

One of the more complicated scenarios to troubleshoot for an AG are lags in secondary replica processing of primary changes. Delays could be caused by resource issues on the primary, network latency, or resource/processing issues on the secondary.

SQL Server includes great diagnostics to help drill into all these possibilities. Start here to build a plan: `https://learn.microsoft.com/sql/database-engine/availability-groups/windows/troubleshoot-primary-changes-not-reflected-on-secondary`.

One of the interesting scenarios that causes a delay is a concept called *flow control.* Flow control is a concept where SQL Server on the primary replica of an AG will delay sending changes to the secondary because the secondary replica cannot process changes from the primary fast enough. The concept is to regulate changes to send to the secondary to avoid overconsuming the secondary. The typical scenario for flow control is an issue with the secondary replica to harden changes to its transaction. In our experience, it is usually due to I/O issues on the secondary, a secondary replica that is underpowered, or some type of CPU consumption issue on the secondary.

The problem is that we discovered that if a network is slow between the primary and secondary—but once the secondary receives changes, it can process them quickly—the primary flow control may be unnecessary.

The communication protocol used between replicas is called **Universal Communication Service** (UCS). Data that is packaged using this protocol is combined into a logical concept called a *boxcar*. Since flow control is like a pause or slowdown mechanism, if the secondary can't keep up with the primary, UCS will temporarily reduce the rate of sending messages to avoid overwhelming it by limiting the number of boxcars to use for transmission. All of these decisions are dynamic, and you have no control over the use of boxcars.

That is until now. With SQL Server 2025, we have a new server configuration option called **max ucs send boxcars**. Increasing this value can help avoid or mitigate flow control that might delay changes to your replicas. We have found this applies more to scenarios where your secondary replica(s) are on slower networks (and that is by design). I recommend you only change the default value if you (1) observe flow control and (2) know your secondary is keeping up when it receives changes.

I asked James indicators of this situation. He told me several perfmon counters exist (such as **Database Flow Control Delay**) and DMVs such as **sys.dm_hadr_database_replica_states** that contains the value called **log_send_queue_size**. This shows you if flow control is causing you delays. On the secondary, check perfmon counters like **Log Bytes Flushed/sec** to gauge how fast changes are hardened on the secondary log. If this number matches your expectations (and hopefully similar to the primary), then flow control may be caused by a slow network. Since flow control may mask the "delay," you will need to use other independent tools to measure expected network speeds. But if you know you are using a "slow" network, see flow control, and the log is getting flushed at expected rates, setting the boxcar value may help your overall sync performance. James and I talked at length on "what number should you use," and only testing is something can help you determine that. Setting the number above the default will not cause harm unless the secondary is not properly keeping up with hardening log changes

You can read more about this configuration option at `https://learn.microsoft.com/sql/database-engine/configure-windows/ucs-flow-control-sp-configure`. This change requires RECONFIGURE and a server restart.

Tuning group commit wait

Every time SQL Server sends data to any resource, whether it be over a network or to a disk, we attempt to follow a simple principle: it is typically better to send a *small* number of *larger* datasets than a large number of smaller datasets. We discovered some time ago that if the workload feeding the primary replica had small transactions, we were sending a large number of small UCS packets to the secondary and found it was inefficient. Thus was born the concept of *group commit*. Starting in SQL Server 2016, SQL Server uses a *ten-millisecond delay* in an attempt to fill Always On Availability Group log blocks with multiple commits before sending them to secondary replicas.

You can observe group commit through insights like the **HADR_GROUP_COMMIT** wait type and the perfmon counters **Group Commit Time** and **Group Commits/sec** and the **Extended Event hadr_log_block_group_commit**.

This worked great for some time until a customer of ours after some very painful debugging showed us that their network latency to the secondary replicas was so fast that group commit was hurting their overall throughput vs. helping. Therefore, we devised some very complex set of trace flags to help. You can see this effort for SQL Server 2017 and 2019 at `https://support.microsoft.com/topic/kb4565944-improvement-a-manual-method-to-set-maximum-group-commit-time-in-sql-server-2017-and-2019-13ef10f5-71d0-5873-0010-f5d1ad4798fe`.

We decided for SQL Server 2025 to make this much more supportable and flexible and so have introduced a new server configuration option called **availability group commit time (ms)**. This option requires a RECONFIGURE but not a server restart. Setting the value of 0 sets the group commit time to the default of 10ms. Any value between 1 and 10 is acceptable.

Should you consider using this? James and I chatted once again and felt this would only help if (1) your workload involves a lot of very small transactions and (2) your network latency is well below 10ms. I tried various small transaction scenarios on my Azure VM setup and never saw any increase in overall throughput rates sending transactions to the secondary.

But your configuration and setup may benefit from this. I didn't find any scenario that lowering this value less than 10 causes any harm. But as with other options listed in this chapter, test it out and see if it can help your AG experience.

Configuring the AG

I have found over my career even what appears to be the smallest of enhancements matters to at least one customer. We have built two improvements based on long-time customer requests to make configuration of an AG easier.

REMOVE listener IP address.

In SQL Server 2025, you can now remove a specific IP address from a listener without deleting the listener.

Set NONE for read-only or read-write routing.

Read-only and read-write routing are an important part of any AG configuration. If you want to revert a specific IP address and port you have set up for the AG, you can now set these options to NONE in SQL Server 2025.

Improved health check timeout diagnostics

Having a central place to see key diagnostics for any problem is important. SQL Server provides many different diagnostics logs and options to see the health of an AG only from a SQL Server perspective. The Windows Cluster Log is the ultimate *source of truth for everything* about the cluster, including the health of the AG. While there are some diagnostics published to the cluster log for a SQL Server health check, we added more details in SQL Server 2025. Our documentation at `https://learn.microsoft.com/sql/database-engine/availability-groups/windows/availability-group-lease-healthcheck-timeout#sql-server-2025-improved-health-check-timeout-diagnostics` calls out an example like the following:

```
ERR   [RES] SQL Server Availability Group: [hadrag] Failure detected,
diagnostics heartbeat is lost
WARN  [RES] SQL Server Availability Group: [hadrag] AG health check failed,
logging perf counter data collected so far
WARN  [RES] SQL Server Availability Group: [hadrag] Date/Time, Processor
time(%), Available memory(bytes), Avg disk read(secs), Avg disk write(secs)
WARN  [RES] SQL Server Availability Group: [hadrag] 4/18/2024 23:55:25.0,
21.857418, 3248349184.000000, 0.000000, 0.000253
WARN  [RES] SQL Server Availability Group: [hadrag] 4/18/2024 23:55:35.0,
11.442071, 3255394304.000000, 0.000907, 0.000382
WARN  [RES] SQL Server Availability Group: [hadrag] 4/18/2024 23:55:45.0,
9.979768, 3253981184.000000, 0.000415, 0.000549
WARN  [RES] SQL Server Availability Group: [hadrag] 4/18/2024 23:55:55.0,
9.762850, 3251232768.000000, 0.001989, 0.000638
WARN  [RES] SQL Server Availability Group: [hadrag] 4/18/2024 23:56:5.0,
9.827234, 3250462720.000000, 0.002250, 0.001418
```

DAG Enhancements

A Distributed Availability Group (DAG) is a critical part of any disaster recovery enterprise critical platform. To get familiar with this concept if you have not used this before, look at our docs at `https://learn.microsoft.com/sql/database-engine/availability-groups/windows/distributed-availability-groups`.

DAG supports contained AGs

Contained Availability Groups (CAGs) were an important enhancement that was added to the SQL Server 2022 lineup of features. The concept if you have not seen this is that the master and msdb system databases become part of the Availability Group. This means any system-level objects like logins and SQL Server Agent jobs can now operate after a failover on the new primary. What really happens behind the scenes is that a separate master and msdb database are created as "user databases" in your AG. But anytime you connect to the AG with the listener (or directly with the context of a database in the AG), any system-level operation like creating a login or SQL Server Agent job is added to these "system databases." And since these databases are part of the AG, changes are automatically applied on the secondary. SQL Server Agent has been enhanced to only execute jobs on the "true primary." You can read more details at `https://learn.microsoft.com/sql/database-engine/availability-groups/windows/contained-availability-groups-overview`.

As soon as we delivered this feature, customers asked us whether CAGs could be supported for a Distributed Availability Group (DAG). SQL Server 2025 now adds this capability. So you can now configure the primary AG of the DAG to be a contained AG, and then you can configure the primary of the second AG (the forwarder) to also be a contained AG. Think of the scenario where you create a SQL Agent job to help maintain statistics using the listener. Now if the primary AG fails over to its secondary, the SQL Server Agent job works without any intervention. In addition, should you fail over to the other AG of the DAG, the SQL Server Agent job also works there.

DAG sync improvement

When you set up a DAG, it is highly recommended that the second AG uses the same *commit mode* as the first AG. Consider Figure 7-20 that I pulled from the documentation (`https://learn.microsoft.com/sql/database-engine/availability-groups/windows/distributed-availability-groups`).

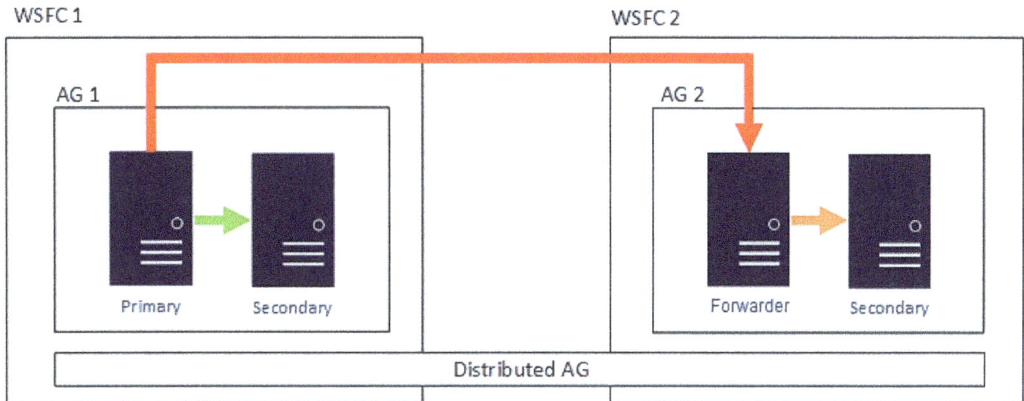

Figure 7-20. *The architecture of a DAG*

In most configurations, AG 1 uses a **synchronous commit mode** to its secondary. Our recommendation is that for AG 2 the forwarder uses the same commit mode to its secondary. But we have seen customers who either misconfigure AG 2 or just want AG 2 to use *asynchronous* commit mode (the orange arrow). The problem here is we assume AG 2 is using synchronous. According to James, *"...basically if customers had a mismatch between GP and forwarder before (which is not recommended), we always sent sync message through the DAG even for async forwarder. Now we honor an async configuration on the forwarder side, meaning we don't view the forwarder as sync anymore. This means if they have a mismatch between GP and forwarder (not recommended), they may have latency in replica LSNs now where they didn't see it before, even though that realistically makes sense."*

It is not technically possible to force these commit modes. So instead, in SQL Server 2025, we honor that AG 2 may be using async and therefore improve performance of synchronization between AG 1 and AG2. This is all built-in when you create a DAG with SQL Server 2025. No special configuration required. However, we stick by our recommendation to use the same commit mode for both AGs in a DAG. Just an example of us reacting to the real world of AG configurations across the world.

Backup/Restore

I can't remember a major release of SQL Server where we didn't add enhancements to backup or restore. And SQL Server 2025 is no different. All of these improvements are based on real customer feedback and innovation: a better compression algorithm, *finally* all backup types on secondary replicas, and backups to an immutable storage location to respond to ransomware threats.

ZSTD Backup Compression

We first introduced the concept of *compressed* backups in SQL Server 2008. The algorithm used by SQL Server from this origin is called **MS_EXPRESS**. MS_EXPRESS is not available through a published API, but the algorithm for how the compression works can be found at https://learn.microsoft.com/openspecs/windows_protocols/ms-xca/a8b7cb0a-92a6-4187-a23b-5e14273b96f8?redirectedfrom=MSDN.

The *compression ratio* for backups that use MS_EXPRESS all depends on what is *compressible* in the database pages and log blocks of the database. I backed up the sample AdventureWorks on SQL Server 2025 using the default algorithm of MS_EXPRESS and achieved an 86% compression ratio. You can see this using the following query:

```
SELECT database_name, backup_size,
compressed_backup_size,
100.0 * (1 - (compressed_backup_size * 1.0 / backup_size)) AS
CompressionPercent
FROM msdb..backupset
WHERE compressed_backup_size > 0;
```

My results were the following:

```
database_name    backup_size    compressed_backup_
size    CompressionPercent
AdventureWorks   4481024          596044                        86.698486774452
```

In SQL Server 2022, we introduced another compression algorithm supported through Intel QuickAssist Technology (QAT) called **QAT_DEFLATE**. This compression technology requires drivers from Intel and when combined with Intel hardware can help offload CPU resources required for compression to the specialized hardware. This algorithm may also give you a better compression ratio depending on the makeup of your data. You can read more about QAT at https://learn.microsoft.com/sql/relational-databases/integrated-acceleration/overview.

In 2016, Yann Collet from Facebook published a new open source compression algorithm called **ZStandard** or **ZSTD** (https://github.com/facebook/zstd). The promise of ZSTD is *better, faster* compression that is **tunable**. In SQL Server 2025, we have now incorporated ZSTD as a third compression option for SQL Server backups.

A new **WITH COMPRESSION** option **ZSTD** offers an additional parameter for level LOW, MEDIUM, or HIGH with LOW being the default. LOW offers the fastest compression with the lowest compression ratio. You can then move to MEDIUM and HIGH, where HIGH sacrifices speed for the best compression ratio.

Note You can use a server configuration option to set what the *default* backup compression algorithm is called **backup compression default**. The only issue with using this option for ZSTD is you can't specify LEVEL, which uses LOW as the default.

Using AdventureWorks again as a sample and queries I showed above with MS_EXPRESS, using a syntax like this

```
BACKUP DATABASE AdventureWorks TO DISK = 'c:\temp\adw.bak' WITH COMPRESSION
(ALGORITHM = ZSTD, LEVEL = LOW), INIT, FORMAT
```

resulted in an 87% compression ratio, so it was just slightly better than MS_EXPRESS, but it ran faster (0.047 seconds for MS_EXPRESS to 0.035 for ZSTD). That time doesn't seem much, but for larger backups it can mean a lot. A level of HIGH resulted in a 90% compression ratio, but it took 0.057 seconds. In addition, the higher the level, the more CPU resources are required, but testing has shown even ZSTD at HIGH is on par with CPU resources required for MS_EXPRESS.

Don't just trust me on this. Anthony Nocentino from Pure Storage (and SQL MVP) wrote an excellent blog with detailed testing on ZSTD at `https://www.nocentino.com/posts/2025-05-27-using-zstd-backup-compression/`. Anthony concluded, *"SQL Server 2025 introduces the ZSTD compression algorithm, a significant advancement in database backup technologies and performance tuning. The default LOW setting offers impressive performance improvements that will be advantageous for most environments. It is recommended to test with your specific workloads to identify the optimal configuration for your needs."*

Backup on Secondaries

When we introduced Always On Availability Groups in SQL Server 2012, we also introduced the concept of executing backups based on the secondary replica. The thinking is you might want to offload backup operations to a secondary instead of running this on the primary.

We also introduced a concept for you to set the preference on how backups should work so you can build automatic backup logic via SQL Server Agent. So you can run a query like this in your job, which would be set up on all replicas:

```
DECLARE @DBNAME NVARCHAR(128) = N'YourDatabaseName';
IF (sys.fn_hadr_backup_is_preferred_replica(@DBNAME) != 1)
BEGIN
    PRINT 'This is not the preferred replica, exiting with success';
    RETURN 0; -- Normal exit, no error
END
BACKUP DATABASE @DBNAME
TO DISK = N'C:\Backups\YourDatabaseName.bak'
WITH COPY_ONLY, COMPRESSION;
```

The **sys.fn_hadr_backup_is_preferred_replica**() function uses a configuration you provide when you create the AG called **AUTOMATED_BACKUP_PREFERENCE**. This value determines if backups should be run on the primary or secondary (there is also an option called **BACKUP_PRIORITY** that affects this decision). You can read more about this topic at `https://learn.microsoft.com/sql/database-engine/availability-groups/windows/configure-backup-on-availability-replicas-sql-server`.

Notice in my T-SQL example I used a COPY_ONLY backup. That is because prior to SQL Server 2025 we only support COPY_ONLY and transaction log backups on secondaries. Starting in SQL Server 2025, we also support full and differential backups. I asked Dinakar Nethi, the lead product manager for this feature, how are we able to now support full and differential backups. He told me it requires code to synchronize certain activities on primary and secondary. For example, he said, *"...It means performing checkpoint on primary and wait for secondary to catch up to the checkpoint by applying log records during database backup."* Having an option to perform all of your backups on a secondary has been a request we have seen from our customers for some time, so I'm glad to see this make the SQL Server 2025 release.

Backup to Immutable Azure Storage

Have you ever heard of a WORM (Write Once Read Many) drive? As far back as the 1980s, we used CDs and eventually optical DVD as worm drives, especially to back up data. You can write once to these drives but only once. This prevents any tampering of the drive. (If you really want to go back in time, tape drives could also be set up this way, but I'm dating myself now.) Back in these days the main reason for WORM drives was to protect the integrity of the data, which was required by certain industry or government standards.

Backing up data to cloud providers has become very popular, and we have enabled SQL Server to back up to Azure Storage via the BACKUP TO URL syntax since SQL Server 2012. Around 2020, Azure started offering *immutable* storage (`https://learn.microsoft.com/azure/storage/blobs/immutable-storage-overview`). The proliferation of *ransomware* attacks is what really drove innovations for immutable storage in Azure.

The problem for SQL Server is our backup algorithm can require multiple write commits to Azure Storage. So, in SQL Server 2025, we changed our algorithm when writing a backup to URL to only commit the writes for the entire backup set one time, thus now allowing our customers to support backups to Azure immutable storage.

Hidden Gems

Every release of SQL Server has a set of features that don't make the big headlines but always have real value. I call these *hidden gems*. There are several in SQL Server 2025 related to the core engine worth calling out to round out the chapter.

ABORT_QUERY_EXECUTION

Consider this situation. You observe a query that is performing poorly, for example, consuming a significant amount of CPU, and is affecting the overall health of SQL Server and other queries. To add to the story, this query is not necessarily required to run your mission-critical application. The problem is you can't get the query changed by the source application and you are in a time-sensitive situation. What if you could within SQL Server "tag" this query to fail anytime it is executed until you can get the situation under control? Welcome to the ABORT_QUERY_EXECUTION query store hint.

The best way to understand how this can work is to use an example.

Prerequisites

All scripts for this example can be found in the **ch7 – Core Engine/ABORT_QUERY_EXECUTION** folder.

1. Download the sample database **AdventureWorks** from `https://github.com/Microsoft/sql-server-samples/releases/download/adventureworks/AdventureWorks2022.bak`.
2. Restore the database using the script **restore_adventureworks.sql** (you may need to edit the file paths for the backup and/or database and log files).
3. Enable the query store by executing the script **enablequerystore.sql**. This will enable the query store for the AdventureWorks database.

The Bad Query

To show you how this works, I needed a "bad query." I thought to myself, *I'm sure Copilot can help me because creating "bad queries" is not in my nature.* I built a prompt for the Copilot preview in SSMS to help me build a "bad query." It was incredibly hilarious that Copilot did not want to do this. No one must have trained AI models to do the wrong thing. It took a while but finally I found one.

1. Finally, after several prompts I was able to get a query based on the **AdventureWorks** sample that looked like the script **poorquery.sql**:

```
USE [AdventureWorks];
GO
WITH LargeDataSet AS (
    SELECT
        p.ProductID, p.Name, p.ProductNumber, p.Color,
        s.SalesOrderID, s.OrderQty, s.UnitPrice, s.LineTotal,
        c.CustomerID, c.AccountNumber,
        (SELECT AVG(UnitPrice) FROM Sales.SalesOrderDetail WHERE
        ProductID = p.ProductID) AS AvgUnitPrice,
```

```
        (SELECT COUNT(*) FROM Sales.SalesOrderDetail WHERE
        ProductID = p.ProductID) AS OrderCount,
        (SELECT SUM(LineTotal) FROM Sales.SalesOrderDetail WHERE
        ProductID = p.ProductID) AS TotalSales,
        (SELECT MAX(OrderDate) FROM Sales.SalesOrderHeader WHERE
        CustomerID = c.CustomerID) AS LastOrderDate,
        r.ReviewCount
    FROM
        Production.Product p
    JOIN
        Sales.SalesOrderDetail s ON p.ProductID = s.ProductID
    JOIN
        Sales.SalesOrderHeader h ON s.SalesOrderID =
        h.SalesOrderID
    JOIN
        Sales.Customer c ON h.CustomerID = c.CustomerID
    JOIN
        (SELECT
             ProductID, COUNT(*) AS ReviewCount
         FROM
             Production.ProductReview
         GROUP BY
             ProductID) r ON p.ProductID = r.ProductID
     CROSS JOIN
       (SELECT TOP 1000 * FROM Sales.SalesOrderDetail) s2
)
SELECT
    ld.ProductID, ld.Name, ld.ProductNumber, ld.Color,
    ld.SalesOrderID, ld.OrderQty, ld.UnitPrice, ld.LineTotal,
    ld.CustomerID, ld.AccountNumber, ld.AvgUnitPrice,
    ld.OrderCount, ld.TotalSales, ld.LastOrderDate, ld.ReviewCount
FROM
    LargeDataSet ld
ORDER BY
    ld.OrderQty DESC, ld.ReviewCount ASC;
GO
```

This query uses a Common Table Expression (CTE) to show a query for "analytics," but it is not well designed.

2. **Run the bad query.**

   Use SSMS to run this query. It takes ~13 seconds resulting in well over 1 million rows! And the query just chews up CPU.

So, bad query, "make my day"

I'm such a movie dork. I love quoting famous lines of movies with my wife and my family for real-world scenarios. And they just roll their eyes and say, "What movie was that from?" Clint Eastwood was in so many iconic movies from my youth. But in the 1980s one of the most famous quotes from his movie *Sudden Impact* was "Make my day." Ok, I'm being overdramatic to use that quote for a query performance problem, but that is what ABORT_QUERY_EXECUTION does provide you. As an admin, you have control to cause a query that is causing disruption to a mission-critical application to fail (or abort) anytime it is executed.

So how can we control this query? With a query store hint.

1. Load the script **findtopdurationqueries.sql** in a SSMS query editor window.

   Execute the query in the script. This will show you the top queries by duration in the query store. The query at the top should be the query in poorquery.sql. Verify the **query_sql_text** matches the query in poorquery.sql. Note down the **query_id** for the query.

2. Load the script **setabortqueryhint.sql** in a SSMS query editor window.

   This script will cancel the query in poorquery.sql. Replace the query_id in the script with the query_id from the previous step. Execute the query in the script.

3. Execute the query again in **poorquery.sql**. You will see that the query is canceled immediately with the following error:

```
Msg 8778, Level 16, State 1, Line 1
Query execution has been aborted because the
ABORT_QUERY_EXECUTION hint was specified
```

4. Optionally you can use the script **clearabortqueryhint.sql** to clear the ABORT_QUERY_HINT for the query. This will allow the query to run normally.

   A very nice feature to keep your system up and running while you resolve the long-term problem.

In-Memory OLTP

In SQL Server 2016 we released a new feature for super-low-latency transactions and in-memory processing called In-Memory OLTP (codename Hekaton, which in Greek means 100 – our goal was to enable 100× faster transaction performance). When you enable In-Memory OLTP, a specific file(s) and file group are created to support In-Memory OLTP transactions. When this is enabled, there is no way to "disable" In-Memory OLTP from the database without dropping the database. Enabling In-Memory OLTP is not just about these special files but also several background tasks (the In-Memory OLTP "engine") that are created along with memory consumers.

Now in SQL Server 2025 you can remove all In-Memory OLTP memory-optimized tables, files, and file groups, thereby removing the "engine" components, without dropping the database. There are a specific set of steps to do this, which you can read at `https://learn.microsoft.com/sql/relational-databases/in-memory-oltp/memory-optimized-container-filegroup-removal`.

Note Thank you, Dimitri Furman, for pushing this to make the release after so many years of it not making the cut!

PolyBase

Data virtualization through a feature called PolyBase has been in the product since SQL Server 2016. In SQL Server 2025, we have made using PolyBase easier and added support for managed identities.

PolyBase Services May Not Be Required

It started with access to Hadoop. Then we added ODBC Driver support. These methods all require the PolyBase Services to be installed and a server configuration option.

In SQL Server 2022, we added *direct* support for Parquet, Delta, and other file formats. Direct means all communication is done *inside* the SQL Server engine and doesn't require PolyBase Services. The only problem is that we didn't have time to "unwire" the dependency for the services to be installed.

So you would run this T-SQL to enable PolyBase:

```
EXEC sp_configure 'polybase enabled', 1;
RECONFIGURE;
```

But you would then get this error:

```
Msg 46923, Level 16, State 1
The PolyBase feature must be installed and enabled to use this statement.
```

You would then need to go back and install the PolyBase feature, which installs PolyBase Services.

Now in SQL Server 2025, to use Parquet, Delta, or other "text" files, you don't need any configuration. Therefore, without any special feature or configuration required, you can run a query like this

```
-- Query on a file of parquet files stored in a publicly available storage
account:
SELECT * FROM OPENROWSET( BULK 'abs://nyctlc@azureopendatastorage.blob.
core.windows.net/yellow/puYear=2001/puMonth=1', FORMAT = 'parquet') AS
taxidata;
GO
```

and get results immediately. This query accesses an Azure Storage account that is public to anyone for testing purposes.

Managed Identity Support

You learned in Chapter 6 how Azure Arc now provides Managed Identity support for SQL Server 2025. In that chapter, I showed you how to use a Managed Identity to secure access to an Azure AI Foundry model instead of an API key. Earlier in this chapter I mentioned other examples that take advantage of Managed Identities including BACKUP TO URL and Azure Key Vault.

PolyBase now supports a Managed Identity to secure access to Azure Blob Storage and Azure Data Lake Storage for files like Parquet and Delta. The beautiful thing about this feature is you configure this very similar to access to AI models like you saw in Chapter 6 using a DATABASE SCOPED CREDENTIAL for the system-assigned Managed Identity. Your system-assigned Managed Identity will need proper permissions to access Azure Storage, which you can read all the details on in our documentation at `https://learn.microsoft.com/sql/relational-databases/polybase/managed-identity`.

Diagnostics

One of the things Dimitri and I are both passionate about is diagnostics. Dimitri along with our team introduced a new Dynamic Management View (DMV) called **sys.dm_os_memory_health_history.** Memory is a very complicated topic for SQL Server. The design of the product is intended so you don't have to know all the details about how it works. But I've seen in situations in my career where customers encounter out-of-memory (OOM) conditions or performance issues and are not sure why. This DMV is intended to help with those situations.

The concept is the engine will capture information about memory "health" in a ring buffer style as a snapshot every 15 seconds for up to 256 snapshots of the current state of memory consumption in the engine. This includes how much memory can be allocated by the engine, how much memory could be reclaimed from caches, the top memory consumers in the engine (in the form of clerks), and status of memory health. Let's look at an example:

Let's assume I have SQL Server 2025 installed using Developer (Enterprise) Edition. I didn't change any of the defaults for max server memory during setup. My laptop has 32GB of RAM, so some of your numbers in this example will vary. I have installed SQL Server but have not added any databases.

1. Let's first check two key metrics for memory—committed and target—from using the script **dm_os_sys_info.sql** that runs the following T-SQL statement:

   ```
   SELECT committed_kb, committed_target_kb
   FROM sys.dm_os_sys_info;
   GO
   ```

On my laptop the results looked like this:

```
committed_kb      committed_target_kb
567504            12561848
```

These results show that at startup SQL Server has committed ~554MB while can grow up memory to ~12GB.

2. Now let's see what the new DMV says from the script **memoryhealth.sql** that has the following T-SQL statement:

```
SELECT * FROM sys.dm_os_memory_health_history;
GO
```

This yields about 85 rows. The first two rows look like Figure 7-21.

	snapshot_time	allocation_potential_memory_mb	reclaimable_cache_memory_mb	top_memory_clerks	severity_level
1	2025-08-03 19:25:21.0033333	30310	17	[{"clerk_type":"MEMORYCLERK_SQLCLR","pages_alloc...	1
2	2025-08-03 19:25:05.9866667	30313	15	[{"clerk_type":"MEMORYCLERK_SQLCLR","pages_alloc...	1

Figure 7-21. *Memory health at server start*

The **allocation_potential_memory_mb** is how much "free" memory SQL Server has to grow its allocations. This is around 30GB. This means SQL Server can grow memory significantly based on available memory on my laptop. The column **reclaimable_cache_memory_mb** is how much memory from caches can be reclaimed since that is the nature of caches like the buffer pool. The **top_memory_clerks** column is a JSON document describing the top five memory consumer components in the engine. The column **severity_level** shows overall memory health. **1** means low indicating good memory health.

3. Let's force a memory pressure situation to see what health could look like under extreme pressure. Execute the script **memorypressure.sql** that runs the following T-SQL statement:

```
sp_configure 'show advanced', 1;
go
reconfigure;
```

```
go
sp_configure 'max server memory', 600;
go
reconfigure;
go
```

4. Check committed and target again using **dm_os_sys_info.sql**. The results now look like

```
committed_kb      committed_target_kb
580216            614400
```

   You can see the target has been lowered to the max server memory value.

5. Check memory health again using **memoryhealth.sql**. The results look like Figure 7-22.

	snapshot_time	allocation_potential_memory_mb	reclaimable_cache_memory_mb	top_memory_clerks	severity_level
1	2025-08-03 19:41:37.4600000	41	17	[{"clerk_type":"MEMORYCLERK_SQLCLR","pages_alloc...	3
2	2025-08-03 19:41:22.4333333	29860	17	[{"clerk_type":"MEMORYCLERK_SQLCLR","pages_alloc...	1

Figure 7-22. *Memory health after pressure*

   Notice the allocation potential is considerably lower and severity_level is now **3**. There is room to reclaim cache but not much. This is a good example of what poor memory health looks like. While this is an external memory pressure situation (other programs taking all the memory on your machine or VM), there are some SQL Server scenarios that can lead to these problems such as excessive memory grants. Here is a very good Microsoft documentation article to get you started: `https://learn.microsoft.com/troubleshoot/sql/database-engine/performance/troubleshoot-memory-issues`.

I believe this DMV can be very useful in the right situations or help you proactively monitor healthy memory usage for SQL Server 2025.

The Fastest Database on the Planet

SQL Server has always had a reputation for performance. It all started with the desire to compete with Oracle with TPC benchmarks all the way back to SQL Server 6.5. Now SQL Server dominates the performance landscape as a standard for database performance. One of the reasons this tradition keeps on is our performance team. Focused purely on making SQL Server the fastest database on the planet, this team, led by Partner Software Engineer Thierry Fevrier, operates on a different level.

For SQL Server 2025, the team invested significant effort well before the launch of Project Kauai to ensure the platform not only scales seamlessly on cutting-edge hardware but is also fully optimized for both Windows Server and Linux environments—despite the extensive new feature set introduced across SQL Server solutions. SQL Server is such a unique workload that often our work helps improve the Windows Server operating itself. This team works directly with customers, our OEM partners, the Windows Server OS team, and partners like Red Hat for Linux. Their work sometimes requires tools to measure SQL Server performance at the CPU instruction level.

I asked Thierry about his experiences during the development of SQL Server 2025: *"Tuning the SQL Server engine is like tuning a high-performance racecar—except it should run smoothly without needing an expert driver. Our team is always exploring how SQL Server can scale with the latest hardware innovations and real-world customer workloads. We work closely with customers and our incredible partners to make that happen. One of the most exciting aspects of optimizing SQL Server performance is the opportunity to influence how the underlying operating system behaves. I've known Dave Cutler for years, and during SQL Server 2025 development, I was able to bring him into some deep optimization discussions with the core Windows Server OS team. The result? A major performance boost not just for SQL Server, but for all Windows Server customers."*

Now that is legacy intersecting with today's innovation to make SQL Server fast!

I asked Thierry how SQL Server customers would see the benefit of these optimizations for SQL Server 2025. He said, *"Our TPC benchmark results will clearly demonstrate the improvements, but many customers should notice better query performance simply by upgrading to SQL Server 2025. One of our key goals was to ensure that customers leveraging the latest processor advancements would see strong scalability with SQL Server 2025."*

I decided to put Thierry to the test, so I ran some basic tests using the famous HammerDB TPROC-H benchmarks against SQL Server 2022 and SQL Server 2025 using Windows Server 2025. I was pleasantly surprised to see the same workload achieve about a 5–8% gain in raw performance with SQL Server 2025! Not all workloads will see this improvement, but it is very nice to see performance enhancements come with the core engine.

Always Tuning the Engine

I hope you can see from this chapter that the SQL Server engine is alive and well. I know when we announced public preview we surprised many in the community and industry of all the new features for the core engine that is in SQL Server 2025. And you can see that many of these features are based on direct customer feedback or experiences.

Naveen Prakash, Vice President for Engineering and long-time SQL Server veteran, summed it up for me, *"SQL Server 2025 continues to push the boundaries of innovation—delivering performance and reliability gains that applications benefit from automatically, with no code changes required. It also introduces long-awaited enhancements across key surface areas while enabling modern AI-powered database applications through advanced support for JSON and vector data."*

Now that you have toured enhancements for the core engine, explore how to integrate your data with the unified data platform, Microsoft Fabric.

CHAPTER 8

Integrating SQL Server 2025 with Microsoft Fabric

I remember when Microsoft Fabric (Project Trident) was announced at the Microsoft Build conference in May 2023. When this announcement happened, I knew then that our analytic integration solution for SQL Server 2022, Synapse Link, was probably going to be short-lived (`https://learn.microsoft.com/azure/synapse-analytics/synapse-link/sql-server-2022-synapse-link`). I was one of the people at Microsoft touting Synapse Link as a solution for near-real-time analytics for SQL Server using the power of Synapse. The technology was sound, and it worked quite well. However, after spending time talking to customers during the launch of SQL Server 2022, I realized that customers were not excited about the requirement for a Synapse dedicated pool. Dedicated pools are great for powerful warehouse workloads, but in this case, our customers really wanted access to the *Landing Zone*. The Landing Zone is a storage location where we place the "raw" files (e.g., Parquet) required to send data to Synapse. In other words, our customers really wanted near-real-time analytics from SQL Server seamlessly into a *data lake*. And thus began the concept of **Fabric Mirroring**.

Note For those veteran SQL Server users wondering why we would use the term *Mirroring* since SQL has had a technology called Database Mirroring, the precursor to Always On Availability Groups, well, we hashed out a bunch of names with the Fabric team, but Mirroring was the one that landed.

In this chapter I'll describe the fundamentals of Microsoft Fabric and how Fabric Mirroring, specifically for SQL Server, is part of that ecosystem. I'll show you how Fabric Mirroring for SQL Server works and why you should consider it as a solution to offload

B. Ward, *SQL Server 2025 Unveiled*, https://doi.org/10.1007/979-8-8688-1847-9_8

read workloads. I will use an example to show the basics to get started. Then I'll finish off the chapter showing you how to go further with Microsoft Fabric bringing together a truly unified data story.

I have examples in this chapter that will require SQL Server 2025, an Azure subscription, and a Fabric *capacity*. I will describe these requirements in more detail in the examples later in this chapter. As with any of the examples in this book, you may elect to just sit back and enjoy the ride seeing how it all works.

An Introduction to Microsoft Fabric

Initially the theme for Fabric was the *unified analytics platform*. Since that time, we added databases such as SQL Database and Cosmos DB making Microsoft Fabric now the *unified data platform*.

Figure 8-1 is one I often use to describe Microsoft Fabric.

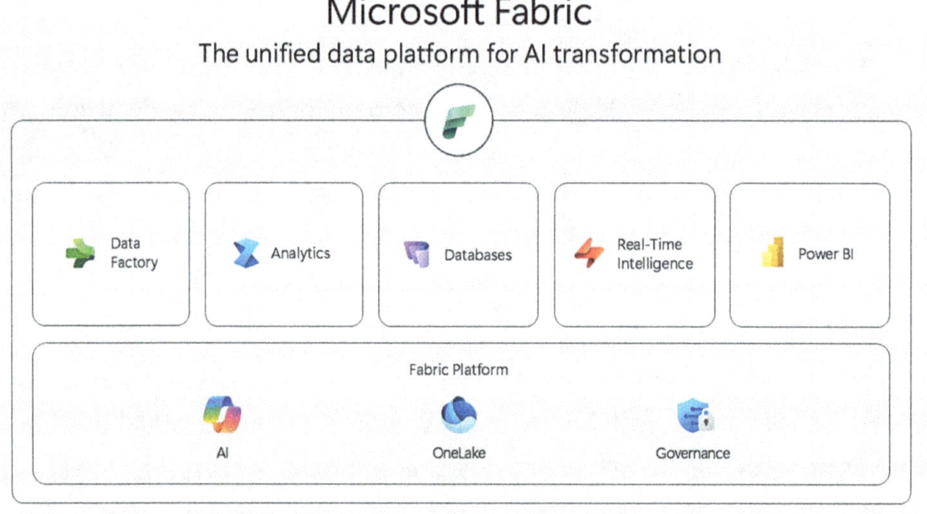

Figure 8-1. *Microsoft Fabric*

If you look at the icons at the top of this visual, you will see both services and solutions including the following.

Data Factory

Services to help you copy, move, and transform data from about any data source to another, both outside and inside Fabric.

Analytics

Services to help you perform analytics on any data you need including warehouses, machine learning, notebooks, and Spark data processing.

Databases

Azure SQL and Cosmos DB running in Fabric? Yes, it is true. The cloud operational databases you know and love now can be deployed within the Fabric experience. I would need to write an entirely new book to cover SQL Database in Fabric (hmmm that is an idea). You can read more about it at `https://aka.ms/sqdbfabric`. I will talk more about it later in this chapter to unify data across Microsoft SQL deployments.

Real-Time Intelligence (RTI)

RTI is a comprehensive, event-driven analytics solution within Microsoft Fabric that enables organizations to ingest, process, analyze, and act on streaming data in real time.

Power BI

Do I need to describe Power BI? The most popular reporting service in the world is known for powerful visuals available in a self-service fashion. The concept of Power BI actually started the thought around Project Trident. I will show you how to use Power BI later in this chapter. But I'm no Guy in a Cube, so if you really want to get deep into Power BI, check out `https://aka.ms/guyinacube`.

These services look great, but what is the magic of Microsoft Fabric? The icons at the bottom explain it (with a few additional opinions of the author).

AI

It is not a surprise that AI is a center of thought for any Microsoft product or service. It is no different for Fabric. Copilot experiences exist everywhere there is a Fabric service. Access to pre-built Azure OpenAI models is simple through AI services (`https://learn.microsoft.com/fabric/data-science/ai-services/ai-services-overview`) and AI functions (`https://learn.microsoft.com/fabric/data-science/ai-functions/overview`). And SQL Database in Fabric supports built-in vector types for vector searching powered by Azure AI Foundry. These are same SQL AI capabilities you learned about in Chapter 4 of the book.

OneLake

For me personally, this is one of the biggest capabilities for Microsoft Fabric and unifies data in a way I've never seen before. I like how Amir Netz describes OneLake. He calls it the "*OneDrive for your data.*" I love OneDrive as a user of Windows. (There is even now a OneLake explorer experience in Windows! (`https://learn.microsoft.com/fabric/onelake/onelake-file-explorer`)). OneLake is the same concept except built for enterprise data. OneLake is a true data lake unifying data across all Fabric experiences. OneLake can be used to store any type of data, but for me its true power comes into play when you store your data in *Delta Parquet* format. You will see in this chapter that this is exactly what Fabric Mirroring does to provide a powerful near-real-time experience.

A part of the OneLake experience is a concept called a *Lakehouse*. I absolutely love the Lakehouse concept, and I'll show you a Lakehouse and how it can help you unify data in a unique way later in this chapter.

Dive into OneLake at `https://learn.microsoft.com/fabric/onelake/onelake-overview`.

Governance

Microsoft Fabric is seamlessly integrated with Microsoft Entra making authentication and permissions easy if you are using that security platform.

In addition, Microsoft Purview is pervasive everywhere in Fabric including Data Catalog, Information Protection, Data Loss Prevention (DLP), and Audit. See more at `https://learn.microsoft.com/fabric/governance/microsoft-purview-fabric`.

In my opinion, there are a few other considerations of why Microsoft Fabric has real value including the following.

Services Working Together

It is easy to work with one Fabric service while using another. For example, I can easily create PySpark notebooks to access OneLake, AI functions, and a new concept called User Data Functions. It is simple to build pipelines and taskflows to orchestrate a data application. Fabric includes an Application Lifecycle Management (ALM) system for CI/CD capabilities across everything you can do in Fabric.

Unified Interfaces

Everything in Fabric is accessible using a common browser experience called the "Fabric portal." User interfaces have consistency across all Fabric services including Copilot chat experiences. As I show you Fabric Mirroring for SQL Server and other examples, you will see what the Fabric user interface looks like.

Fabric also supports a concept called **workspaces**. Workspaces for me are like Azure resource groups. They provide a method to organize your resources and provide a security granularity allowing you to segment off various users and projects. Learn more at `https://learn.microsoft.com/fabric/fundamentals/workspaces`.

Capacity Model

All Fabric services are billed using a consistent concept called the Fabric Capacity Model billed in Fabric *Capacity Units* (CUs). The CU model is a shared model in which any use of resources in your Fabric tenant goes toward consumption of Capacity Units. I will discuss how CUs are related to your use of Fabric Mirroring later in the next section of this chapter. Learning the CU concept can take some time, so I recommend you start with this blog post: `https://blog.fabric.microsoft.com/en-US/blog/fabric-capacities-everything-you-need-to-know-about-whats-new-and-whats-coming`. And then keep up with the latest at `https://learn.microsoft.com/fabric/admin/capacity-settings?tabs=power-bi-premium#manage-capacity`.

Our Fabric teams do an excellent job at publishing roadmaps, so stay in touch with all the latest at `https://aka.ms/fabricroadmap`.

Mirroring SQL Everywhere to Fabric

Now on to the story of Fabric Mirroring. The Fabric team defines **Fabric Mirroring** as "...a low-cost and low-latency solution to bring data from various systems together into a single analytics platform" (`https://learn.microsoft.com/fabric/database/mirrored-database/overview`). After SQL Server 2022, we knew we had the code in both Azure SQL and SQL Server to capture data, including changes from the transaction log (without using CDC), and place them in a Landing Zone (which was Azure Storage) to support the Mirroring concept via Synapse Link.

The Fabric team at the same time was building a *replication* system (a.k.a. Mirroring Services) within Fabric that could take data from a Landing Zone and migrate this to Delta Parquet format into OneLake. The first announcement of Fabric Mirroring came up in November 2023 at Microsoft Ignite ironically from our new product leader of SQL, Priya Sathy: `https://blog.fabric.microsoft.com/blog/introducing-mirroring-in-microsoft-fabric`.

Note Before Mirroring came along, I had already built an example of how to integrate SQL Server data based on Parquet and data virtualization with Fabric shortcuts. I used a cold partition switch scheme to show the example. You can see this on this *Data Exposed* episode: `https://youtu.be/wi4Hg2MDQAY`.

Let's Start with Azure

We started as a private preview for Azure SQL Database. Every time I would talk to customers about Mirroring for Azure SQL, they would ask, "What about SQL Server?" We knew the interest was huge. In March 2024, I did a presentation with Idris Motiwala and Anna Hoffman at the famous Fabric Community Conference on Mirroring. We had a packed room of 500 people. We polled our audience, and everyone wanted Mirroring for SQL Server.

Why would they want this? Because many, like me, saw the promise of OneLake. And many of these users wanted to offload their read workloads off SQL Server but not have to set up a replication subscriber or Always On secondary replica. In addition, they saw the promise of getting data into OneLake to integrate with other data sources and Fabric services.

Mirroring SQL Everywhere

In the spring of 2024, SQL Server 2025 plans were not baked yet. Furthermore, we also wanted to support older versions of SQL Server. Therefore, we started down the path of polishing the architecture from Synapse Link, making some improvements, and working with the Fabric team to connect with their replication system. So we started with Azure SQL Database.

We moved to launch an official public preview for Mirroring for Azure SQL Database in the summer of 2024. And then by the time we announced SQL Server 2025 at Microsoft Ignite in November of 2024, we had Mirroring everywhere SQL existed as seen in Figure 8-2.

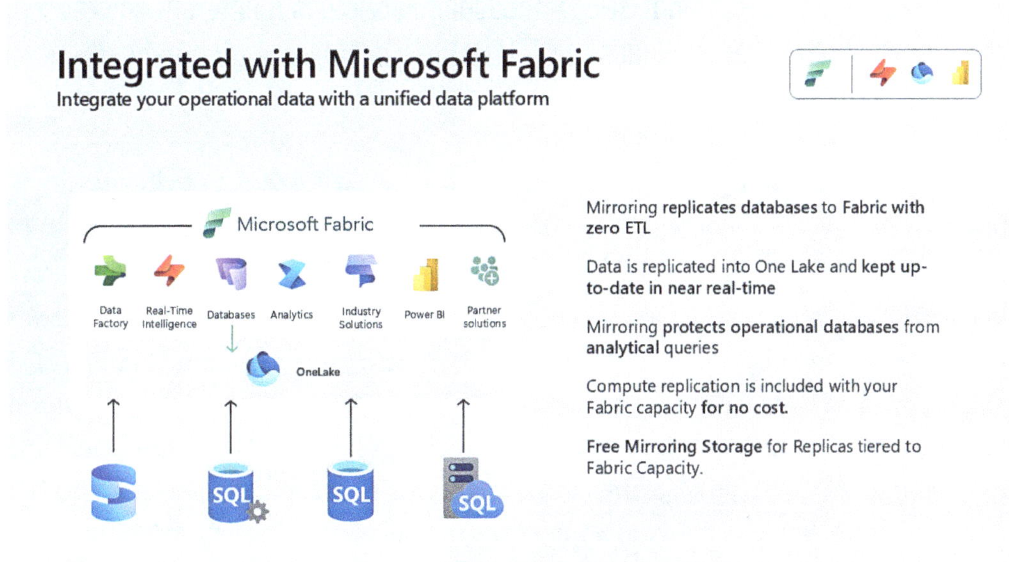

Figure 8-2. *Mirroring everywhere for SQL*

By the time we announced the public preview for SQL Server 2025 in May 2025, we had Mirroring support for SQL as follows:

Azure SQL Database – General Availability using an improved version of the change feed system we had for Synapse Link.

Azure SQL Managed Instance – Public preview using the same change feed system.

SQL Server 2025 - Public preview using the same change feed system as for Azure SQL.

SQL Server 2016-2022 - Public preview using Change Data Capture (CDC) built into SQL Server. Even though we knew some customers may consider using CDC a problem, it is the only technology common to SQL Server 2016-2022. We didn't support the change feed system we built for Synapse Link for SQL Server 2022 because we enhanced it for SQL Server 2025 and didn't want to support the previous implementation.

Note Notice in this visual that SQL Database in Fabric has a different colored arrow into OneLake. That is because SQL Database in Fabric is *automatically* mirrored into OneLake.

One thing you will notice on this visual is this statement:

Free Mirroring Storage for Replicas tiered to Fabric Capacity.

What this means is that you are not charged for the storage for your mirrored data in OneLake per your Fabric capacity (1TB per CU). Fabric capacities come in units of numbers. So, if you have a capacity of unit of F32, you get 32TB of free storage! Anytime you try to query the mirrored data, then you incur charges per the Fabric Capacity Unit shared pricing model (`https://azure.microsoft.com/pricing/details/microsoft-fabric/#pricing`). You can read more about the overall concept of Fabric Mirroring for Azure at `https://aka.ms/sqlmirroring`.

To summarize why someone would want to consider Fabric Mirroring for any SQL deployment, it is to seamlessly copy and synchronize SQL data, without writing any code, into another platform, OneLake, for the purposes of offloading read (analytical) queries but also to integrate this data with other data sources in OneLake.

This book is about SQL Server 2025, so let's use the next section of the chapter to explore the details of how Mirroring for SQL Server 2025 works and use an example to really bring home the story.

Mirroring SQL Server 2025 to Fabric

In this section of the chapter, you will learn the basics of how SQL Server 2025 Mirroring works for Fabric including components, architecture, requirements, and limits. Then what better way to understanding Mirroring than to see it in action.

How It Works

Figure 8-3 is a great way to explain how Fabric Mirroring for SQL Server 2025 works.

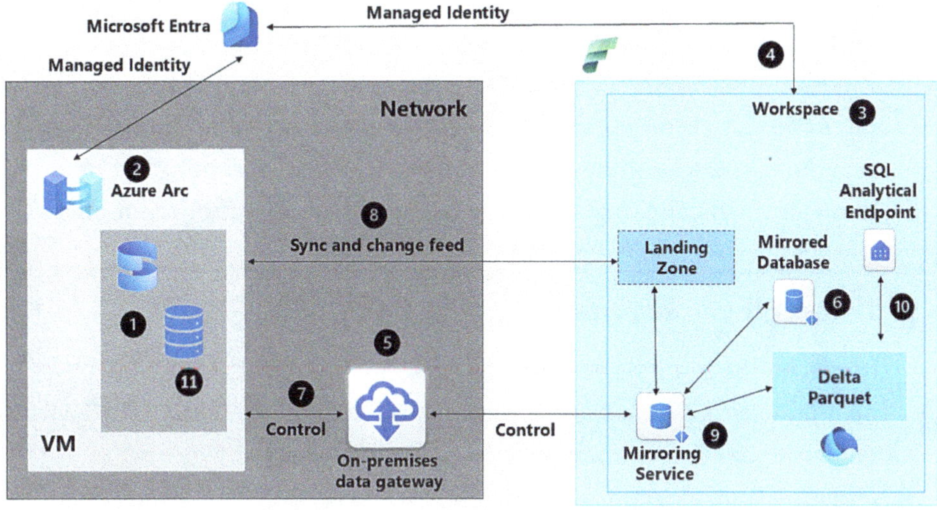

Figure 8-3. Fabric Mirroring for SQL Server 2025 architecture

The numbers in the diagram help explains the components and flow.

1. **SQL Server 2025.**

   You need to deploy SQL Server 2025 and have a database ready to mirror. Mirroring is at the database level, and you can choose which tables (or the entire database) you want to mirror. Later in this section of the chapter, you will read more about limits that might prevent you from using Mirroring for the database or specific tables.

2. **Azure Arc and Managed Identity.**

   You need to enable Azure Arc with Microsoft Entra for SQL Server 2025 and create a system-assigned Managed Identity. This Managed Identity will give SQL Server the right permissions to send data into the Landing Zone.

3. **Create a Fabric workspace.**

   Create a new Fabric tenant and a workspace or in your existing tenant create a new workspace.

4. **Assign the Managed Identity to the workspace.**

   Give permissions to the Fabric workspace with the Managed Identity. Since this is a system-assigned Managed Identity for the Azure Arc resource, the SQL Server engine has permissions to write into the Landing Zone in the workspace. This should be done automatically unless you are using a secondary replica for an Always On Availability Group.

5. **Deploy the on-premises data gateway (OPDG).**

   Install on the same network (or in the same VM or machine) the on-premises data gateway. This gateway, which is also used with Power BI, has been enhanced to support SQL Server Mirroring. OPDG will act as a control plane for SQL Server 2025 Mirroring.

Note For SQL Server 2016–2022, OPDG will also be the data plane as a CDC consumer to send data to Fabric. This means OPDG will also be a "CDC application" to pull the data from SQL Server.

---

6. **Create a mirrored database.**

   Create a mirrored database in Fabric in your workspace. You will also connect using details from the OPDG to SQL Server 2025, choose your tables, and start the Mirroring process.

7. **Mirroring control.**

By starting the Mirroring process, a Mirroring service (there are more details here, but let's just call it a service for now) communicates to the OPDG. The gateway acts a control plane by using system procedures in SQL Server to enable Mirroring for selected tables and providing details of the Landing Zone. This will initiate a sync of the selected tables.

8. **Sync data to the Landing Zone.**

   The SQL Server engine uses a background process to synchronize data from selected tables and send this to the Landing Zone in Fabric.

9. **Replicate to OneLake.**

   The Mirroring service detects the synchronized data in the Landing Zone. It will replicate and convert this data into Delta Parquet files in OneLake in the context of the mirrored database and workspace.

10. **Query using the SQL Analytics Endpoint.**

    Users can now query and explore their mirrored data using the SQL Analytics Endpoint. The SQL Analytics Endpoint is TDS compatible, supports a subset of the T-SQL syntax, and is read-only for the mirrored data. Read more about the SQL Analytics Endpoint at `https://learn.microsoft.com/fabric/database/sql/sql-analytics-endpoint`.

11. **Changes automatically sent to Fabric.**

    With this data synchronized, you can make changes in the SQL Server 2025 database and have them automatically reflected in near real time in Fabric OneLake. Background tasks will pick up committed transactions in the transaction log and send them to the Landing Zone automatically. The Mirroring service picks these up and updates the Delta Parquet files.

Note This is why Delta Parquet is such a good format. It supports transactions against Parquet files. Technically Fabric uses a concept called *Delta verti-parquet*. You can read more about this at `https://learn.microsoft.com/fabric/data-engineering/delta-optimization-and-v-order`.

At this point, you can just let the system run on its own and use the mirrored data in Fabric. You can also use Fabric to stop or start the mirror. The Mirroring service will pick this up and communicate with the OPDG to control SQL Server to stop or start Mirroring again.

Mirroring for Azure SQL Database and Managed Instance works similar to this architecture except you don't need a gateway. As I started earlier, SQL Server 2016–2022 requires Change Data Capture (CDC) to be enabled on the database. The OPDG is responsible for not just control but to send data as a CDC consumer.

Considerations and Limits

Here are some considerations for using Fabric Mirroring for SQL Server.

Transactions

Mirroring only sends committed transactions after the initial data is synchronized. Transactions are never held up, but transaction log truncation could be affected depending on latency of data stored in Fabric.

What Can Be Mirrored?

Only tables can be mirrored but not views. There are specific limits on what can be mirrored in tables, which you can read more about at `https://learn.microsoft.com/fabric/database/mirrored-database/sql-server-limitations#table-level`. There are also some limits on data types and columns, which you can read about at `https://learn.microsoft.com/fabric/database/mirrored-database/sql-server-limitations#column-level`.

Considerations

At the time of the writing of this book, Fabric Mirroring for SQL Server 2025 is in public preview. Some of the following limits may change even after General Availability. Keep up with the latest at `https://learn.microsoft.com/fabric/database/mirrored-database/sql-server-limitations`. For now, consider these:

- Only the primary database is supported in an Always On Availability Group.
- You cannot use Mirroring if the database is enabled for CDC, Change Event Streaming (CES), or Replication.
- The maximum number of tables in a database that can be mirrored is 500. There are no limits on the size of the database.
- Security features like Row-Level Security, Object-Level Permissions, and Dynamic Data Masking are not propagated to the mirrored database in Fabric. OneLake has its own permission system, which you can use.
- DDL operations are supported on source tables except for partition operations and altering of the primary key. If you perform any DDL, the table must be re-synchronized to the mirror.
- In SQL Server 2025, we are also supporting the ability for you to create a Resource Governor resource pool to manage resources used for Fabric Mirroring.

Example: Mirror a SQL Server 2025 Database to Fabric

What better way to bring home how Mirroring works than to show you an example.

In this scenario we will use a *portion* of the AdventureWorksLT sample database in SQL Server 2025. We will mirror this to Fabric and then run queries to view the data. Then we will make changes in the source database and ensure those changes appear in Fabric. To get started let's look at the prerequisites.

All scripts can be found in the **ch8 – Fabric Mirroring** folder.

Prerequisites

1. **Install SQL Server 2025.**

   Since I wrote this book during public preview, I downloaded SQL Server 2025 and used Enterprise Developer Edition. The final decision on what editions will support Mirroring will be made when the product becomes generally available. I'm using the same SQL Server 2025 instance I used in Chapter 6, so Azure Arc is enabled, and Microsoft Entra is configured. If you have done this, you can skip step #3 below.

2. **Restore the AdventureWorksLT database.**

   Download the AdventureWorksLT sample database from `https://github.com/Microsoft/sql-server-samples/releases/download/adventureworks/AdventureWorksLT2022.bak`. Use the script **restoreadwlt.sql** to restore the database to your SQL Server deployment. You may need to change the locations for files. I used my default Windows admin account to do this as it will make it far easier to connect with the gateway. However, you can also use either a SQL login/user or a Microsoft Entra login/user. Follow these steps at `https://learn.microsoft.com/fabric/database/mirrored-database/sql-server-tutorial?tabs=sql2025#prerequisites`.

Important Because of the use of some user-defined data types (UDTs) in the AdventureWorksLT database, you will need to run the script **alteradwltsql2025.sql** before proceeding. This script changes the columns for user-defined data types into their native types. UDTs are not currently supported for Mirroring.

---

3. **Configure Azure Arc and Microsoft Entra** to get a system-assigned Managed Identity.

   If you have not already done this when you installed SQL Server 2025, follow the steps at `https://learn.microsoft.com/en-us/sql/sql-server/azure-arc/connect`. Then configure Microsoft

Entra, which will also set up a system-assigned Managed Identity by following the steps at `https://learn.microsoft.com/sql/sql-server/azure-arc/managed-identity`.

4. **Deploy a Fabric capacity** if you don't have one already.

   If you don't have a Fabric capacity in your organization, you can use a free Fabric trial at `https://aka.ms/fabrictrial`.

5. **Create a new Fabric workspace.**

   Log in to `https://app.fabric.microsoft.com` and create a new workspace. If you have never created a new workspace, consult the documentation at `https://learn.microsoft.com/fabric/fundamentals/create-workspaces`. I called mine **adwspace**.

6. Assign the **Managed Identity permission** to the workspace.

   Use the settings for the workspace to assign **Contributor** rights to your workspace for the system-assigned Managed Identity for the Azure Arc resource (which is the name of the VM). This step is optional and should only be required for secondary replicas of Always On Availability Groups.

7. **Set up the on-premises data gateway.**

   I will set up the gateway on the same virtual machine as SQL Server, but this can be deployed anywhere on your network, which can connect to SQL Server.

   a. Download the latest version of the gateway from `https://www.microsoft.com/en-us/download/details.aspx?id=53127&msockid=0448b52333796d6425f3a0b332c36cba`.

   b. Execute the installer from the download and install according to these steps: `https://learn.microsoft.com/en-us/data-integration/gateway/service-gateway-install`.

   Use the same email address as your Fabric login when prompted. Use any gateway name you like. You will specify this later in this example.

When my gateway finished installing, I got a screen like in Figure 8-4.

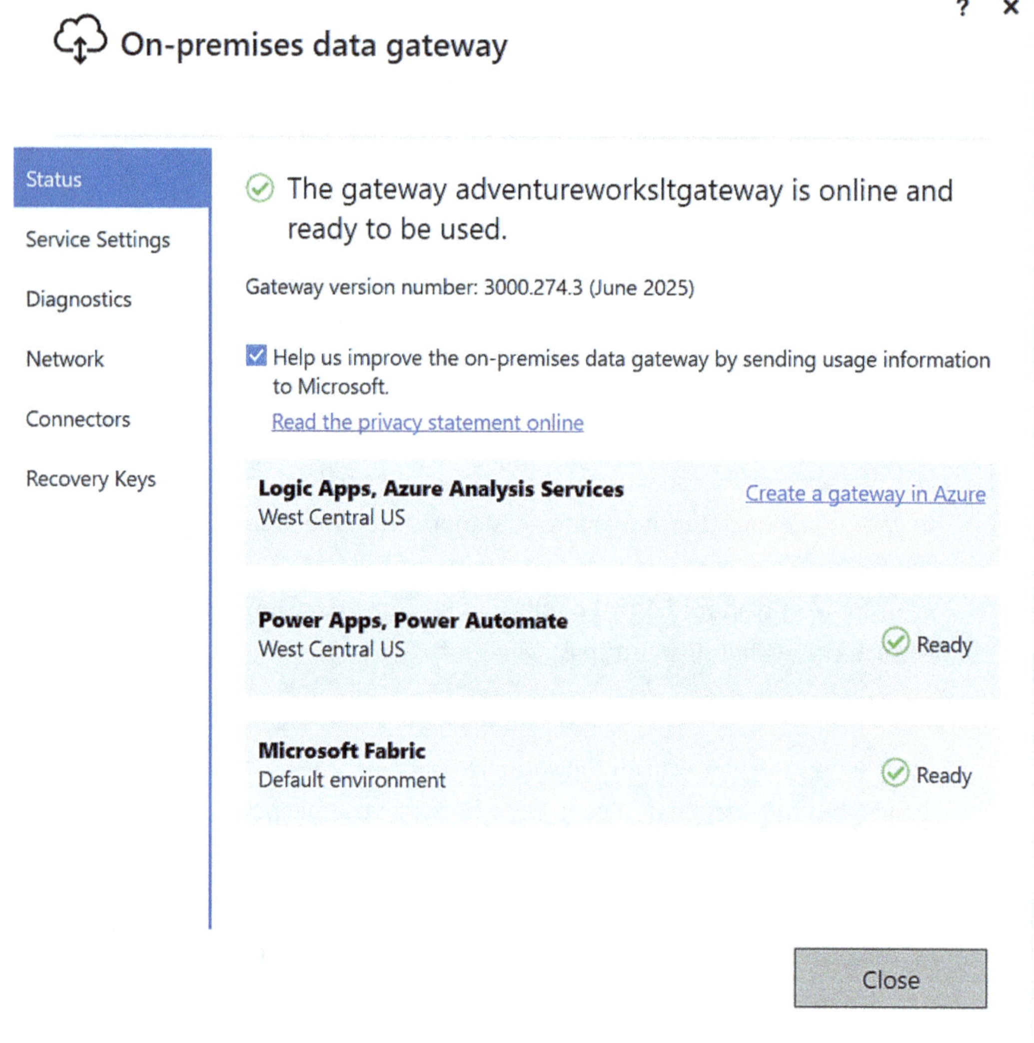

Figure 8-4. *On-premises data gateway successfully installed*

With these prereqs completed, let's create the mirrored database.

Create the Mirrored Database in Fabric

Let's go through the steps to create the mirrored database. When this section is complete, you should have your data synchronized from SQL Server 2025 into Fabric.

1. **Create the mirror** in the Fabric portal.

   In the context of the workspace you created earlier, select + **New Item** and scroll until you see the section **Get Data** and then the option to **Mirror SQL Server** as you can see in Figure 8-5.

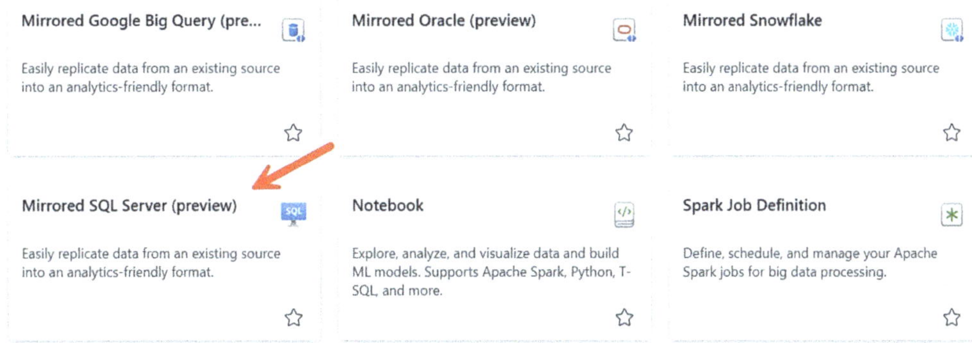

Figure 8-5. *Add a new item as Mirror SQL Server in the Fabric portal*

   You will now be asked to choose a connection. Select + **New** from the left-hand side of the screen and then choose SQL Server Database.

   You now must complete a screen for the connection like in Figure 8-6.

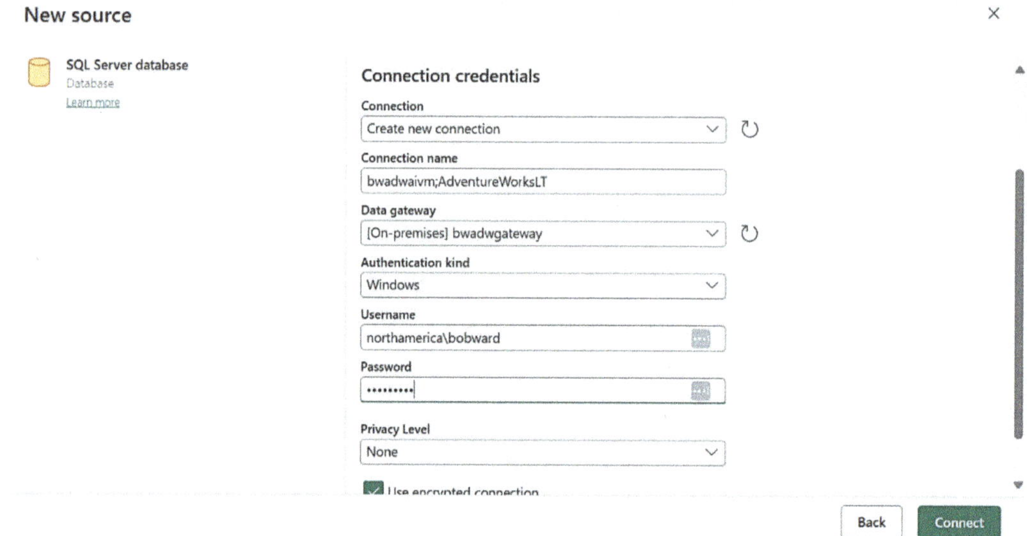

Figure 8-6. *Setting up a new connection for Mirroring*

This doesn't show all the fields, so I'll list out what I selected and choices I made. This is the connection the on-premises data gateway uses to connect to your SQL Server:

Server - I used **.** for local since the gateway is running on the same VM and I'm using a default instance.

Database - The name of the database I'm going to mirror.

When I fill out the Server and Database, the Connection and Connection name are populated automatically.

Data gateway - This is a drop-down list of gateways. The gateways associated with the email address I used when installing the gateway and the one I'm logged into Fabric will appear. I chose the gateway I just deployed.

Authentication kind - I chose the Windows account I used to restore the AdventureWorksLT database because I already have all the permissions required as dbo. However, if you chose a different login/user in the previous steps, use that account.

Privacy Level - I left this to None as this is not used for Mirroring.

You need to uncheck the Use encrypted connection if you are relying on the SQL Server self-signed certificate.

I then selected **Connect**.

Now you can select the tables you want to mirror. In this example, we are only going to choose the **Customer**, **CustomerAddress**, and **Address** tables.

You can make this selection as you see in Figure 8-7.

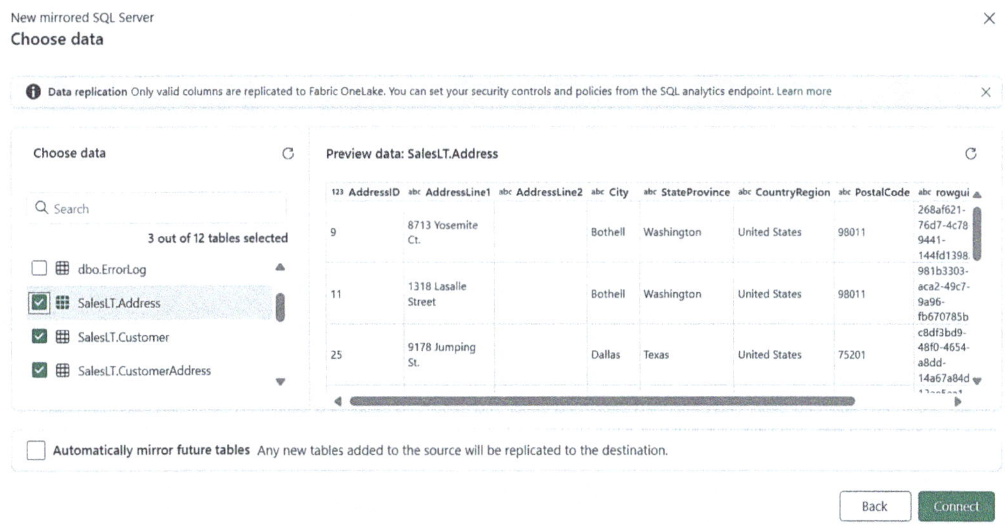

***Figure 8-7.** Choosing tables to mirror*

You can come back later and add more tables if you like. In addition, you could select the option that any new tables created in the future are automatically mirrored. Select Connect to start the Mirroring process.

2. **Start the Mirroring process.**

   By selecting Connect, you have initiated the process to mirror the selected tables. First, give your destination a name. This is what the "database" name will appear in the Fabric interfaces. Because my database will just focus on customers, I called it **AdventureWorksCustomers**. Then select **Create Mirrored Database**.

3. **Monitor until complete**.

   Now you can monitor the status of the mirror from the Fabric interface. Initially, you will see a page like Figure 8-8.

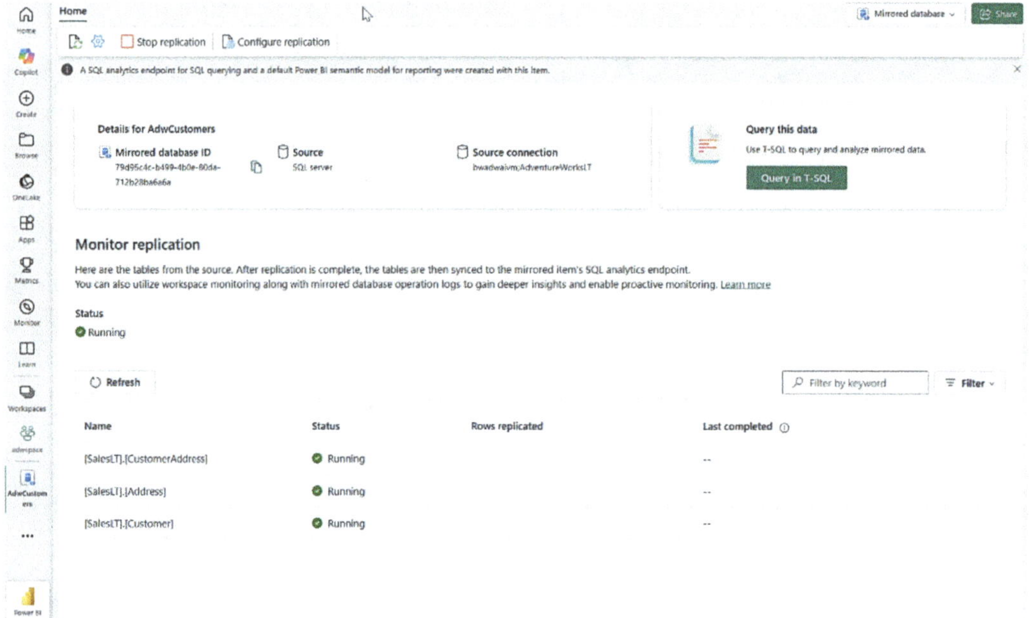

***Figure 8-8.** Monitoring Fabric Mirroring for SQL Server 2025*

The latency of completing the initial sync is based on several factors:

- The size of the tables to synchronize
- The network latency between SQL Server and Microsoft Fabric
- The workloads running on SQL Server (CPU, I/O, etc.) that could affect the ability to run background threads to pull the data and send to Microsoft Fabric

  When the sync is done, your page should look something like Figure 8-9.

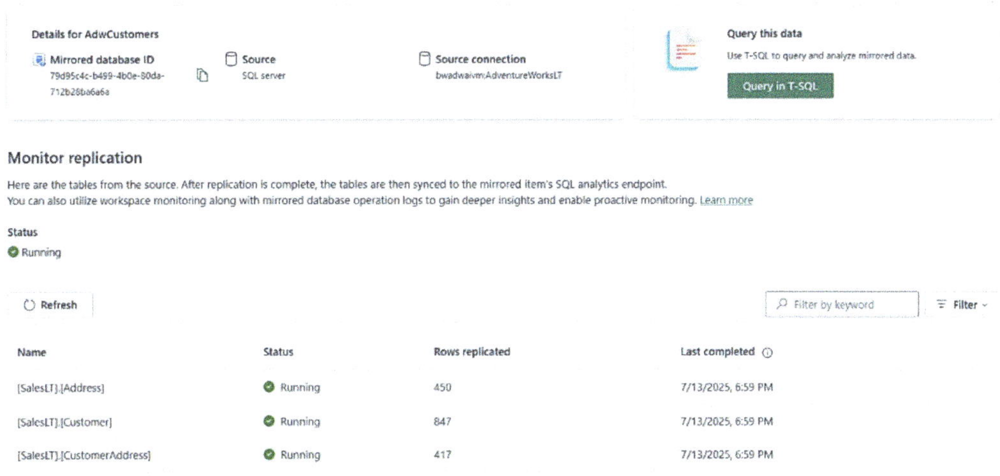

Details for AdwCustomers

Mirrored database ID 79d95c4c-b499-4b0e-80da-712b28ba6a6a

Source SQL server

Source connection bwadwaivm;AdventureWorksLT

Query this data

Use T-SQL to query and analyze mirrored data.

Query in T-SQL

Monitor replication

Here are the tables from the source. After replication is complete, the tables are then synced to the mirrored item's SQL analytics endpoint.
You can also utilize workspace monitoring along with mirrored database operation logs to gain deeper insights and enable proactive monitoring. Learn more

Status

Running

Refresh

Filter by keyword

Filter

Name	Status	Rows replicated	Last completed
[SalesLT].[Address]	Running	450	7/13/2025, 6:59 PM
[SalesLT].[Customer]	Running	847	7/13/2025, 6:59 PM
[SalesLT].[CustomerAddress]	Running	417	7/13/2025, 6:59 PM

***Figure 8-9.** Initial sync completed for the Fabric Mirror*

Let's go to SQL Server 2025 to see what the status of the mirror looks like from the SQL Server perspective.

4. View the **status of the mirror in SQL Server 2025**.

   Connected to SQL Server 2025, run the script **checkmirrorstatus.sql** that uses the following T-SQL statements for Dynamic Management Views (DMV) with SQL Server:

```
USE AdventureWorksLT;
GO
EXEC sp_help_change_feed;
GO
SELECT * FROM sys.dm_change_feed_log_scan_sessions;
GO
SELECT * FROM sys.dm_change_feed_errors;
GO
```

The first result is the configuration of the mirror. The second result is a separate row for the status of the mirror for each table. The third result is blank, but if you have problems, this DMV can be used to troubleshoot problems.

Let's go back to the Fabric portal to view our mirrored data.

Using the SQL Analytics Endpoint

Now we can use the SQL Analytics Endpoint in the Fabric portal to view the mirrored data.

1. **Connect to the endpoint in the workspace.**

   Find your workspace and you will see several objects. Click the SQL Analytics Endpoint you can see like in Figure 8-10.

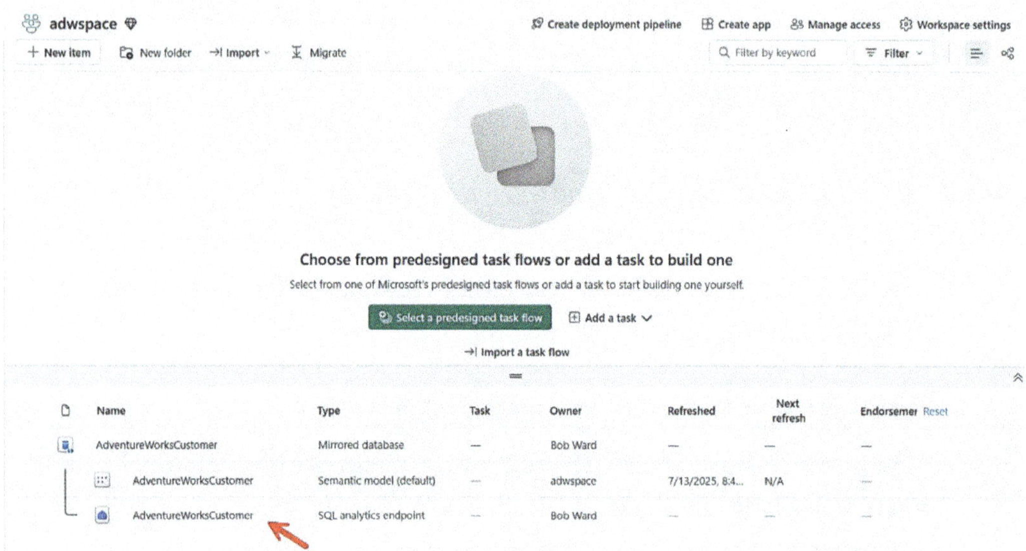

***Figure 8-10.** Finding the SQL Analytics Endpoint for mirrored data*

2. **Explore your mirrored data.**

   After selecting the SQL Analytics Endpoint, your page should look like Figure 8-11 (I expanded the folders to show the tables that were mirrored, which has kept the SalesLT schema from the source database).

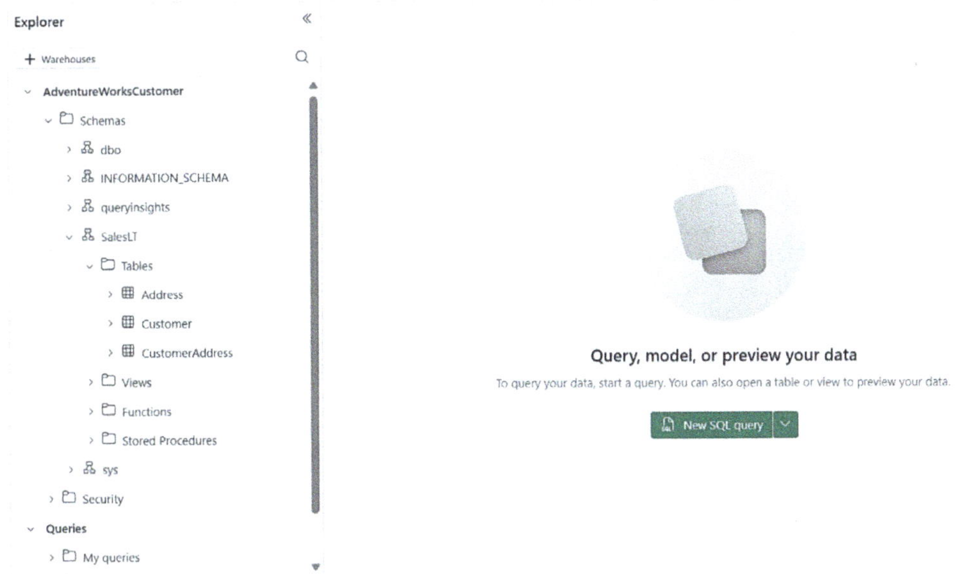

***Figure 8-11.** The SQL Analytics Endpoint experience*

You can see on the left-hand side an "explorer" experience much like you can see with other tools like SQL Server Management Studio (SSMS).

There are several things you can do at this point, but I just want to run a query, so let's use the Copilot experience to help us query the data through joins.

3. **Use Copilot to execute a query.**

   The SQL Analytics Endpoint has a nice query editor, but I'm going to go right to the Copilot experience to help me build a query. Copilot has knowledge of the context of the schema of these tables in the mirrored database.
   I'll first select Copilot in the Fabric UI and put it in a prompt as you can see in Figure 8-12.

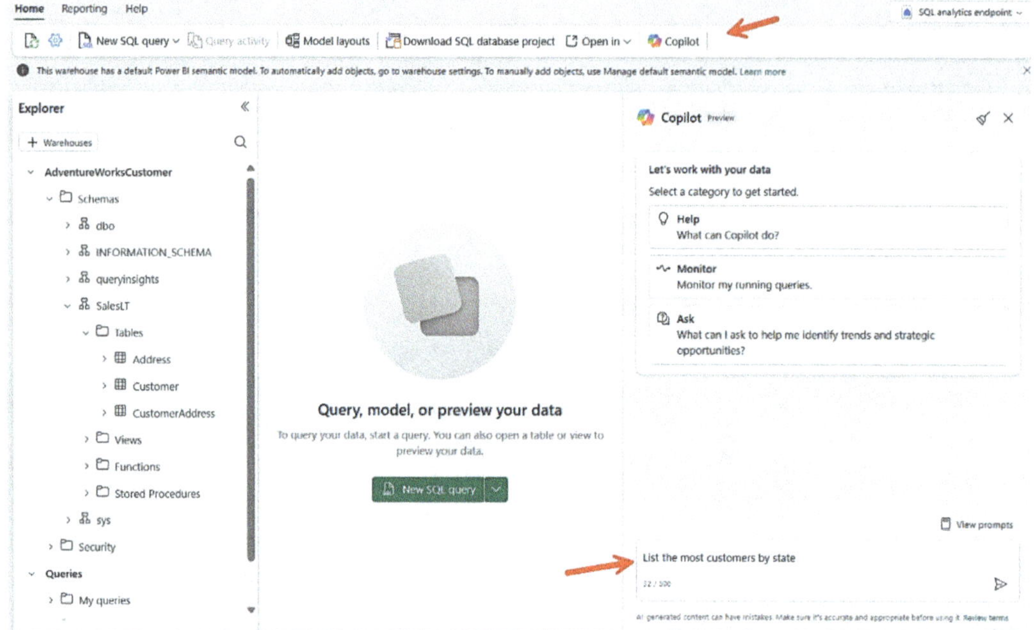

Figure 8-12. *Using a Copilot prompt to create a SQL query*

I chose this prompt because the mirrored data only contains customers and their addresses.

Figure 8-13 shows the generated query and results in the query editor.

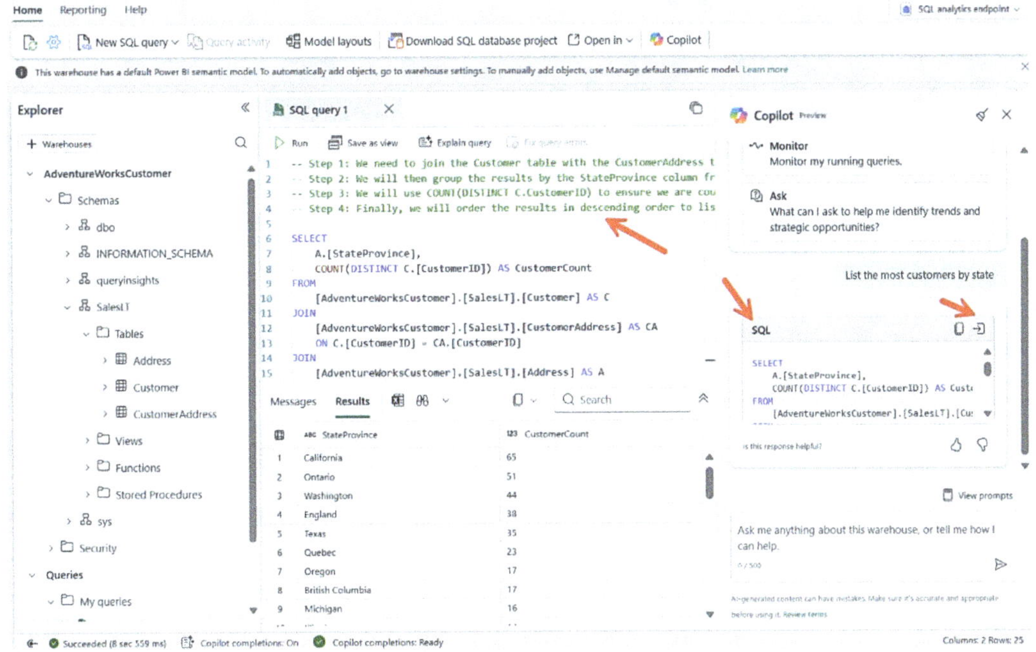

Figure 8-13. *Copilot-generated query for a mirrored database*

Notice the query that is generated uses a *database name* of the name you used to create the mirror. This is effectively the "database name." But all of this is queried against the OneLake mirror of Delta Parquet files, so this is a logical name. Keep this fact in mind as I show you later in the chapter the power of a Lakehouse when we can use a single logical database.

Tip The SQL Analytics Endpoint is a TDS-compatible endpoint. Therefore, you can use a tool like SSMS to connect to it. But what is the connection string? Figure 8-14 shows you how to find it.

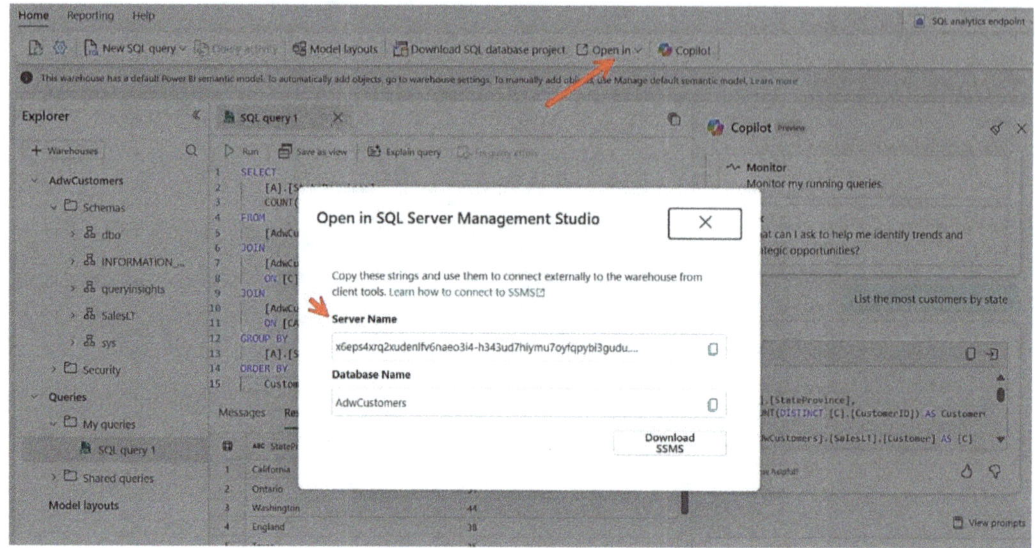

Figure 8-14. *Finding the connection string for SSMS for the SQL Analytics Endpoint*

Now that you know how to access the mirrored data, let's see how you can verify a change to the SQL Server 2025 source database will automatically appear in the mirrored data.

Verify Changes from SQL Server 2025 in Fabric

Any change from the source tables of the mirror should show up in the mirrored database in "near real time" (I always think seconds/minutes vs. hours). Let's run through a few steps to verify this.

1. **Insert a new row in the source database.**

   Use the script **insertnewaddress.sql** to insert a new row into the SalesLT.Address table. This script uses the following T-SQL statement:

   ```
   USE AdventureWorksLT;
   GO
   INSERT INTO SalesLT.Address (
       AddressLine1,
       AddressLine2,
   ```

```
        City,
        StateProvince,
        CountryRegion,
        PostalCode
    )
    VALUES (
        '129 W 81st St, Apt 5A',          -- Jerry's apartment
        'Newman!',                -- Classic Newman moment
        'New York',
        'NY',
        'US',
        '10024'
    );
    GO
```

2. Go back to Fabric and **run a query to see the new row** is there.

   Using the SQL Analytics Endpoint experience, you can click the SalesLT.Address table to see the new appearance like in Figure 8-15.

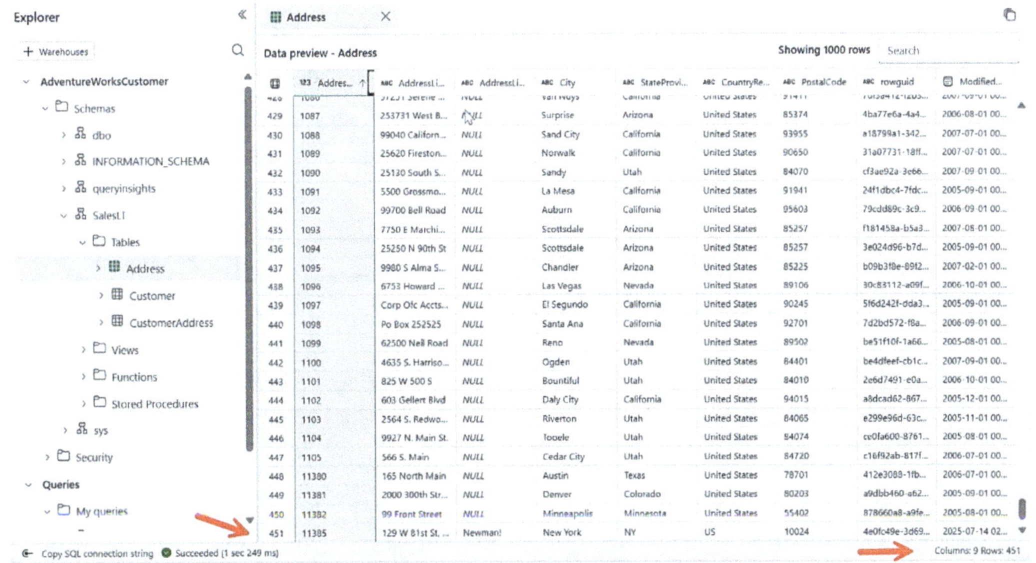

Figure 8-15. *Viewing a new row mirrored from SQL Server 2025*

You can see from the figure the new row added and the row count is one more than the original sync.

The SQL Analytics Endpoint has other capabilities you can use. Besides read-only access to this data using a subset of T-SQL, you have other features to create views and even procedures. This data is stored in a "mini warehouse" but is specific to your mirrored database.

Note You can't mirror other SQL Server objects like views, procedures, and functions—just tables and data.

Now that you have successfully mirrored your SQL Server 2025 database in Fabric, let's go further and see some of the true power of the unified data platform that is Microsoft Fabric.

Going Further with Microsoft Fabric

Let's say that while your customer data for AdventureWorks is in SQL Server 2025, other related data exists in separate databases in SQL Database in Fabric and Azure SQL Database. How cool would it be if you could run queries across all this data in a unified place, with a unified experience and unified query language, and easily integrate this into reports and AI Agents? Well, it sounds cool to me, and that is what I will show you in this part of the chapter. Consider these examples "bonus" steps to go through. If you like, you can just sit back and watch the story unfold as I'll include descriptions and visuals along the way.

Mirroring Azure SQL Database

We have already mirrored the Address, Customer, and CustomerAddress tables from SQL Server 2025. Let's now deploy an Azure SQL Database using the sample option, which is the AdventureWorksLT database. Then we can only mirror the "product" tables into Fabric in the same workspace as above.

I deployed an Azure SQL Database and used these steps from the documentation to mirror an Azure SQL Database: `https://learn.microsoft.com/fabric/database/mirrored-database/azure-sql-database-tutorial`.

I only chose these tables to mirror:

- Product
- ProductCategory
- ProductDescription
- ProductModel

In order to mirror these tables, you will need to connect to the Azure SQL Database and run the script **alteradwltazureqsl.sql** to convert user-defined data types to their native types.

I named the new mirrored database **AdventureWorksProducts**.

Deploying SQL Database in Fabric

The third piece of this story requires you to deploy a SQL Database in Fabric. This is the easiest process for anything I've shown you so far.

Just follow the steps in this tutorial to create the new database: `https://learn.microsoft.com/fabric/database/sql/tutorial-create-database`. I recommend you call it a name like **AdventureWorksSales**.

Then use this documentation to load "sample data," which is the AdventureWorksLT database: `https://learn.microsoft.com/fabric/database/sql/tutorial-ingest-data`.

Now to make the story more real, we need to drop the tables in the database that are not related to "sales." Use the query editor in Fabric and execute the T-SQL statements in the script **alteradwltsqldbfabric.sql**.

Unifying the Data with the SQL Analytics Endpoint

Since SQL Database in Fabric is automatically mirrored, at this point you should have three separate mirrored databases but related due to the original nature of the AdventureWorksLT sample.

Figure 8-16 now shows all my databases, two of which have been mirrored.

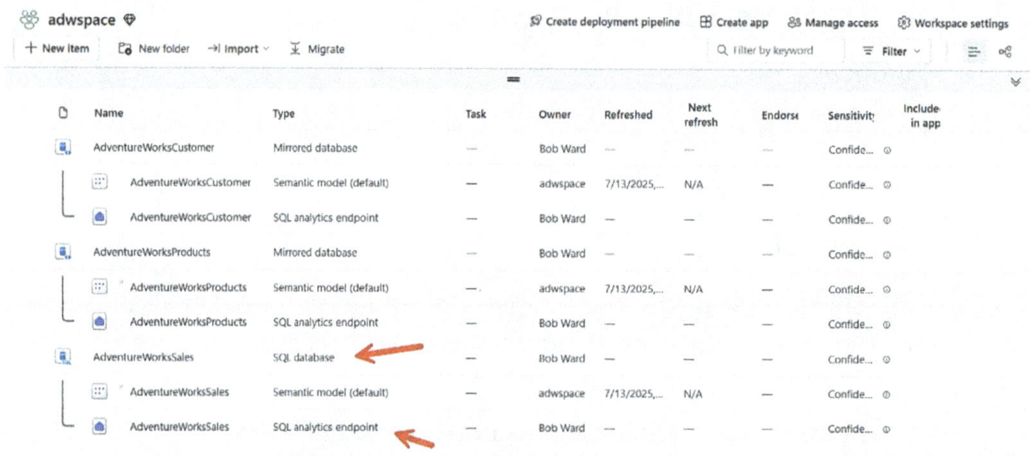

Figure 8-16. *All databases now in the workspace*

Notice that the first two databases are mirrored. You can go into their properties to find their source. The third database is a SQL Database in Fabric, which means it is not mirrored but a ***native*** SQL Database in Fabric. Also notice there is a SQL Analytics Endpoint associated with the database. That is the mirror of the SQL Database in Fabric, which is automatically configured for you.

If you click that endpoint, you will see in Figure 8-17 the sales data for this mirror but also the ability to + **Warehouses** to your view.

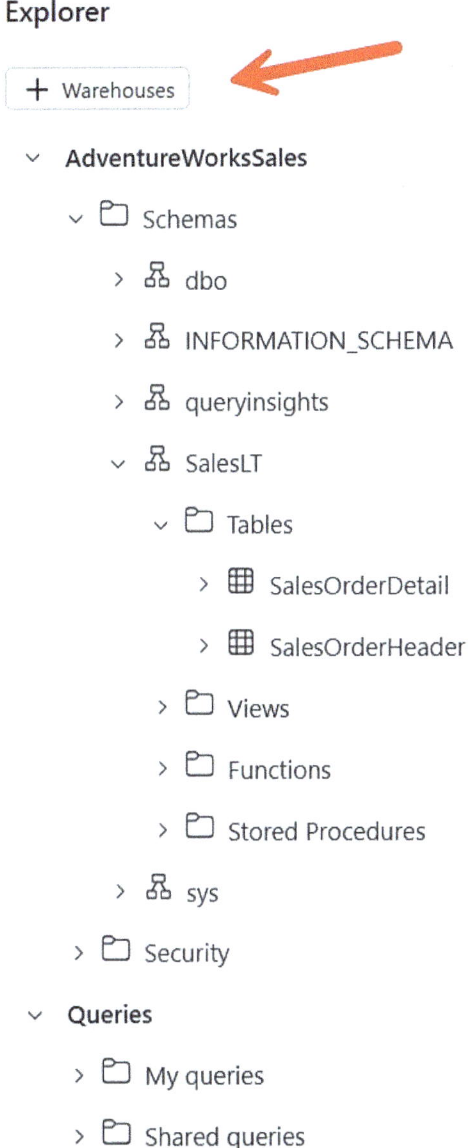

Figure 8-17. *Add a warehouse to the SQL Analytics Endpoint view*

This could be the actual Fabric warehouse in my workspace or the other mirrored databases.

I added the endpoints for both the other mirrored databases, and now my view looks like Figure 8-18.

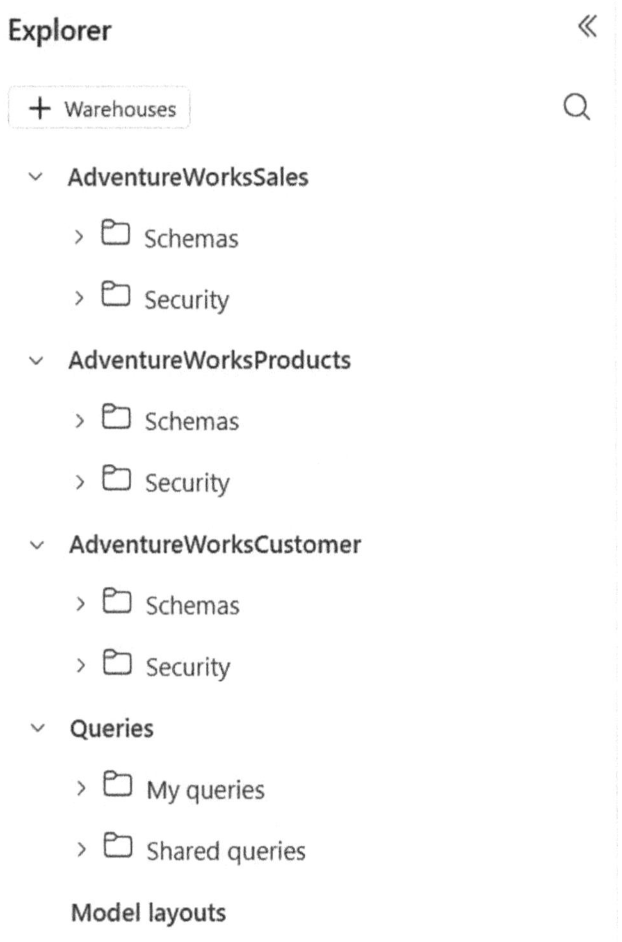

Figure 8-18. *All databases viewable from the SQL Analytics Endpoint*

I can now join this data together, but I will need to reference each database as part of the join. Wouldn't it be nice to see all this data together as one logical database? That is where a Lakehouse comes into play.

Creating a Lakehouse

A Lakehouse in Microsoft Fabric is a logical view over other data sources in OneLake. We will be able to create a new Lakehouse that uses **shortcuts** to our mirrored databases and the SQL Database in Fabric.

If I go back to the context of my workspace, I can select **+ New Item** and then scroll down to the **Store Data** options and pick **Lakehouse**. I will use the name **AdventureWorks** (but not choose the Lakehouse schemas option).

Unified Data in the Lakehouse

Now I'd like to add data to my Lakehouse. This is not a copy of data but a ***shortcut***, which is a link to the original data stored in OneLake.

Figure 8-19 shows how I can choose an option to add a shortcut to my Lakehouse.

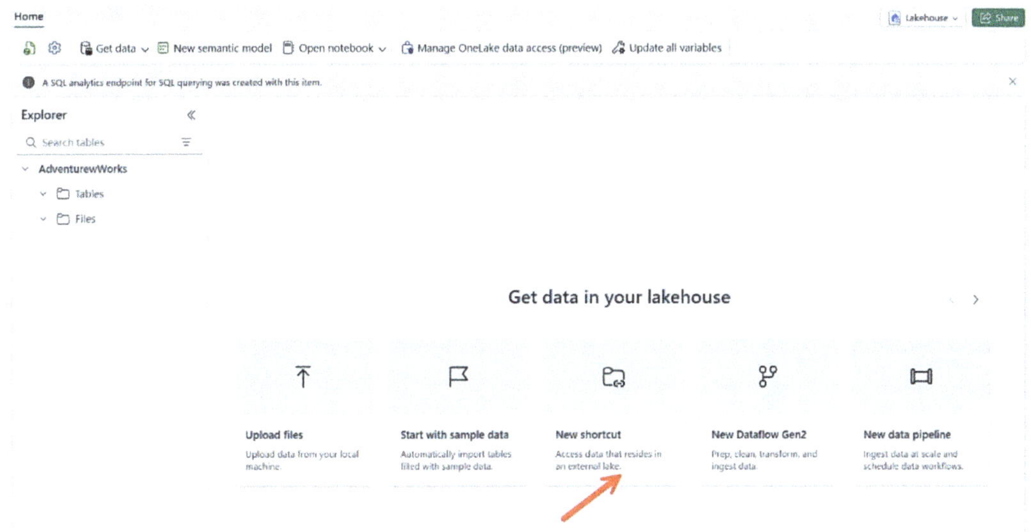

Figure 8-19. *Adding data to a Lakehouse through a shortcut*

I'm presented with a page to choose various data sources. I'll choose OneLake since our data is already there.

You can see in Figure 8-20 all my choices from my workspace.

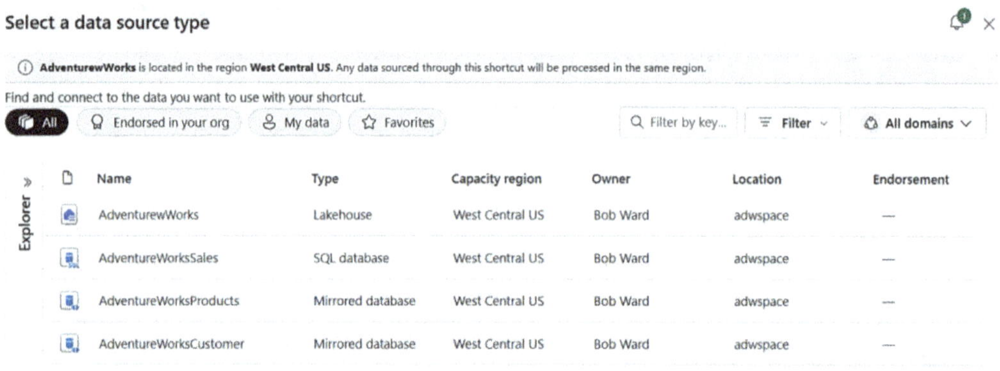

Select a data source type

ⓘ **AdventurewWorks** is located in the region **West Central US**. Any data sourced through this shortcut will be processed in the same region.

Find and connect to the data you want to use with your shortcut.

All | Endorsed in your org | My data | Favorites | Filter by key... | Filter | All domains

Name	Type	Capacity region	Owner	Location	Endorsement
AdventurewWorks	Lakehouse	West Central US	Bob Ward	adwspace	—
AdventureWorksSales	SQL database	West Central US	Bob Ward	adwspace	—
AdventureWorksProducts	Mirrored database	West Central US	Bob Ward	adwspace	—
AdventureWorksCustomer	Mirrored database	West Central US	Bob Ward	adwspace	—

Figure 8-20. *Choosing shortcuts from our databases in the workspace*

I chose each of these databases and hit Next. Unfortunately, you must do these one at a time. Choose each table (don't pick the schema), hit Next, and Create.

When I was done my Lakehouse view looks like Figure 8-21.

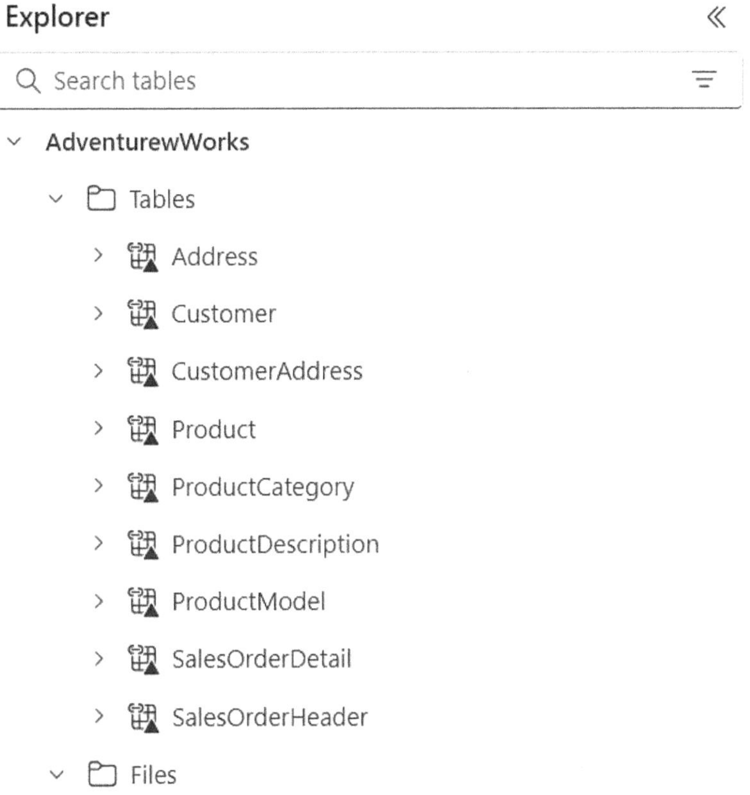

Figure 8-21. *Unified data in the Lakehouse*

How beautiful is this? Notice the icon next to the tables represents they are shortcuts, not actual tables. Any data change to the source is automatically reflected in the shortcut.

Since now the Lakehouse is one single database, I should be able to easily join across of these tables as one logical database. Turns out the Lakehouse also has a SQL Analytics Endpoint. Figure 8-22 shows how you can use Copilot in the SQL Analytics Endpoint to use a prompt to easily join across this data. Copilot knows the schema of these logical tables and the tables are related so can create a join like the following.

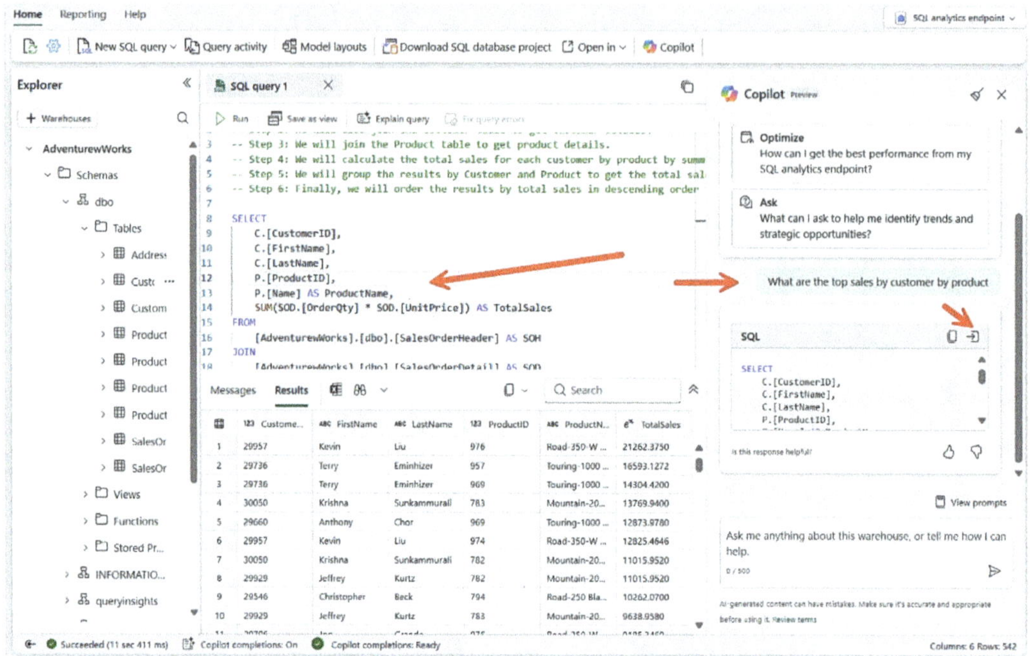

***Figure 8-22.** Using Copilot to join across the tables in the Lakehouse*

Now you are empowered to use a single logical database that comes from three different sources. As a user of the Lakehouse, you don't need to know the details of where the data comes from.

Let's take the next step to take advantage of the power of Microsoft Fabric by creating a Power BI report.

I'm No Guy in a Cube

The promise of Power BI is that it is a self-service platform for building powerful visuals of data. Since I now have a single logical Lakehouse, I can easily build a Power BI report. I am no Guy in a Cube, but I've learned from them that I need a **Semantic Model.**

Fortunately, Microsoft Fabric provides an easy way to build a new Semantic Model on top of the Lakehouse I created.

Using the context of the Lakehouse, I can select New Semantic Model as seen in Figure 8-23.

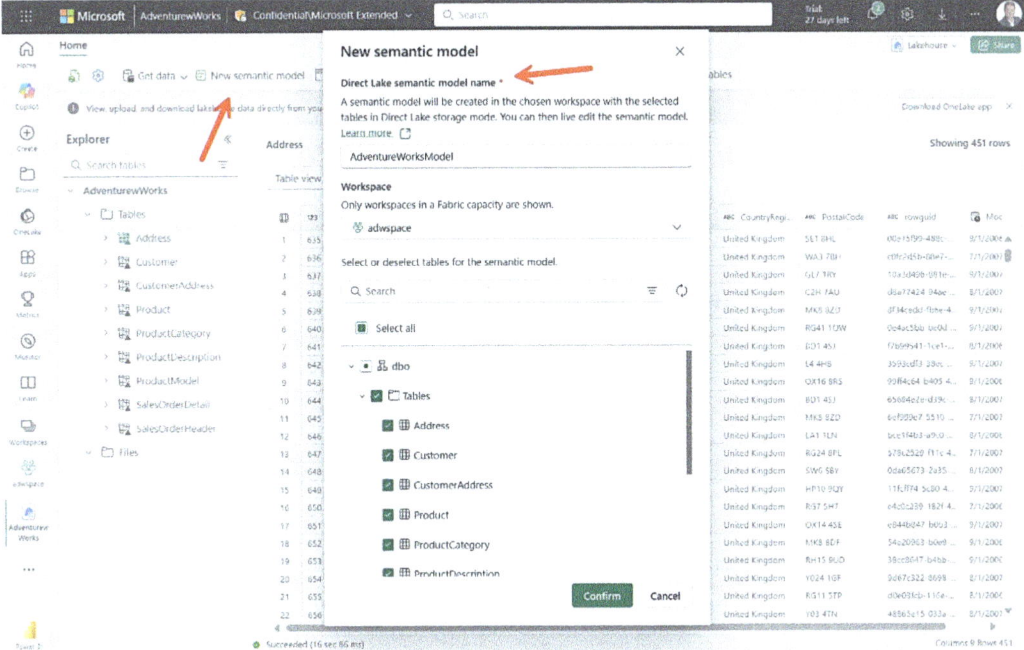

Figure 8-23. *Creating a new DirectLake Semantic Model*

I chose all the tables to create the new model. DirectLake is significant here because using Power BI with DirectLake has a big performance advantage. Learn more at `https://learn.microsoft.com/fabric/fundamentals/direct-lake-overview`.

With the model created, I can click this model from my workspace and as with Figure 8-24 create a new blank report based on this model.

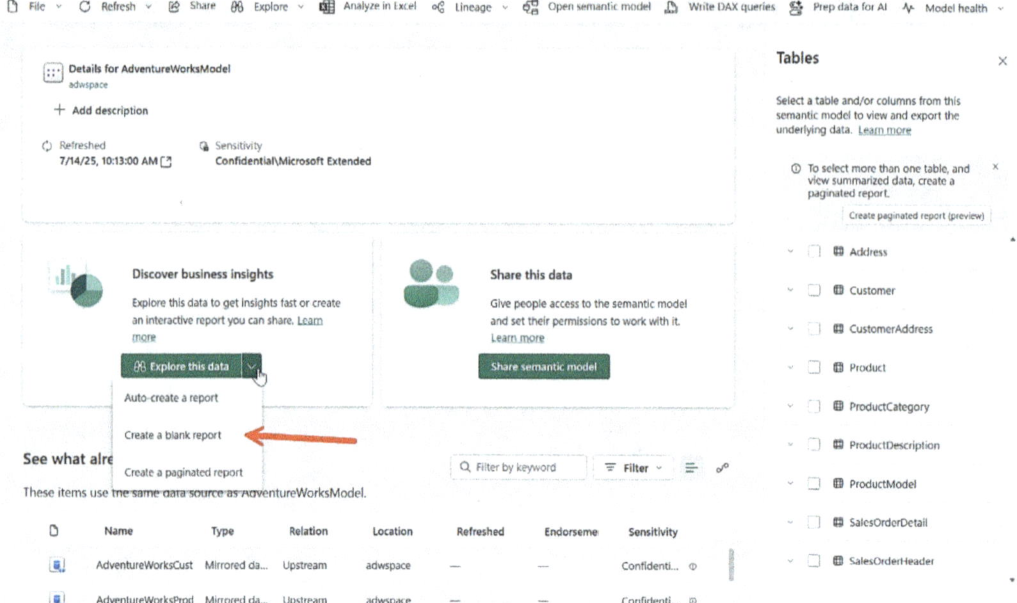

***Figure 8-24.** Creating a blank report from a Semantic Model*

Now with this blank report, I'll use Copilot in Power BI to create a report based on suggested content. Figure 8-25 shows a basic report for product analysis.

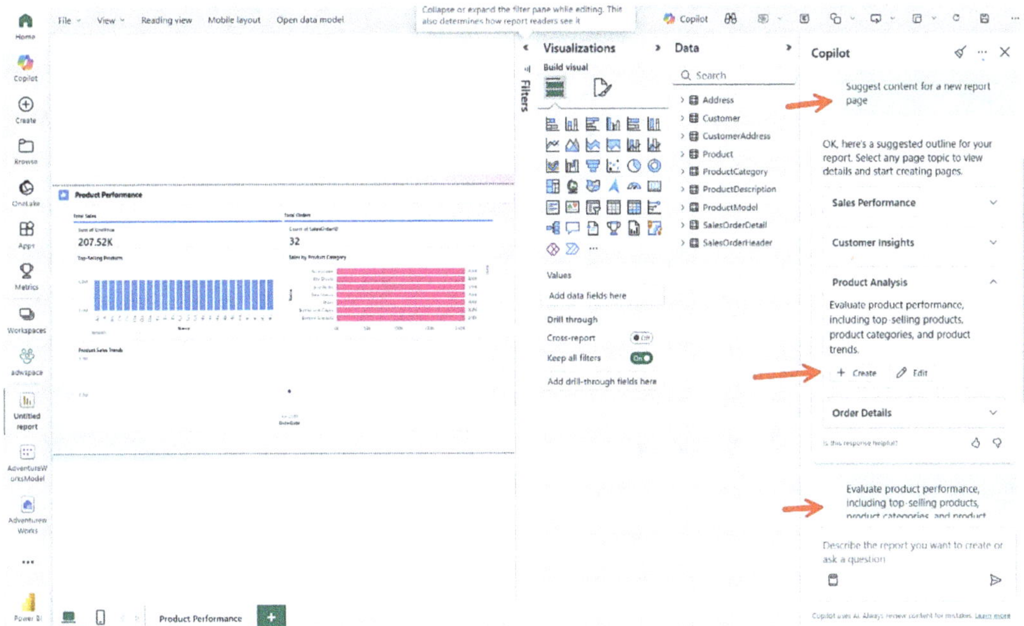

Figure 8-25. *Using Copilot to build a Power BI report*

You can now use Power BI to customize your report. Use the Copilot chat to keep editing the report.

Tip Check out the Power BI **narrative visual** to insert a *Copilot* inside your report. Learn more at `https://learn.microsoft.com/power-bi/create-reports/copilot-create-narrative`.

Is there more? Well, there is a lot more, but one thing I encourage you to look at is Data Agents.

Using Fabric Data Agents

Fabric Data Agents provide the ability to build an AI Agent against Fabric data sources such as a Lakehouse. Figure 8-26 shows how I created a new Data Agent in my workspace and provided a sample chat prompt (notice this is the same prompt I used with Copilot in the SQL Analytics Endpoint) to "chat with my data."

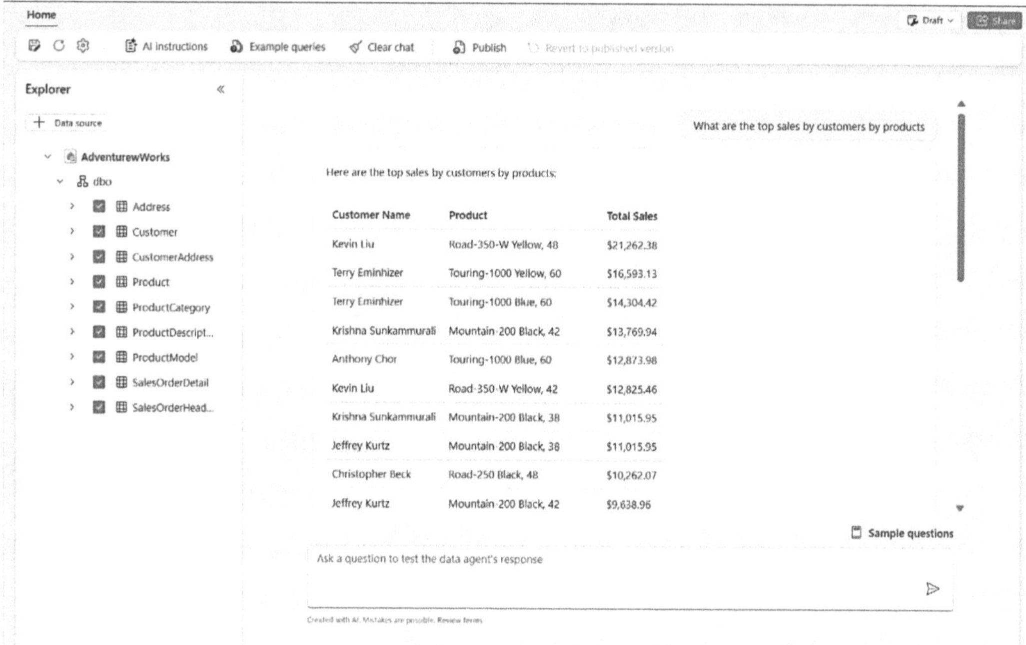

Figure 8-26. *Chatting with a Lakehouse using Fabric Data Agents*

You can do a lot more with Fabric Data Agents including publishing this as an AI application. Learn more at `https://learn.microsoft.com/fabric/data-science/how-to-create-data-agent`.

Good 'Ol SSMS to the Rescue

The final piece of the puzzle to the Fabric Mirroring story is to see how you can connect to your Lakehouse with the most popular SQL tool in the world, SQL Server Management Studio (SSMS). I've mentioned earlier that the SQL Analytics Endpoint is TDS compatible. I also showed you earlier in the chapter how to find the right connection string for the endpoint to use with SSMS.

Microsoft Entra is the only authentication method to access your Lakehouse, so when you use the connection string with SSMS, you need to log in with a Microsoft Entra account that has access to the Lakehouse in Fabric (note you don't need to put in the database context when you connect). Figure 8-27 shows a SSMS connection to the Lakehouse and a query that is the same one generated by Copilot earlier in the chapter.

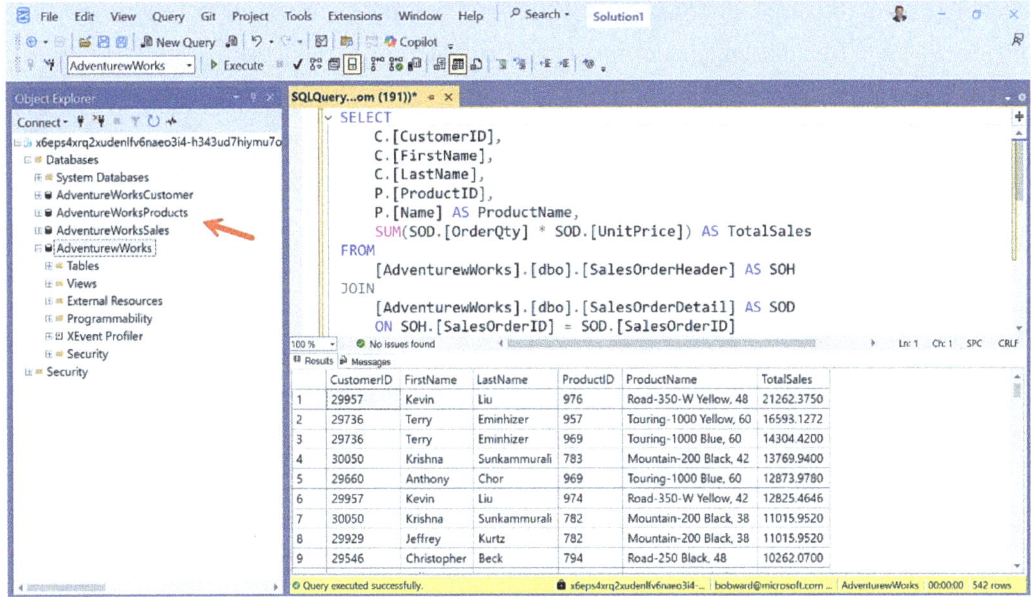

***Figure 8-27.** Running a query with SSMS against the Lakehouse*

Notice that with this endpoint you can query any of the SQL Analytics Endpoints in the workspace. In this case, I ran my query against the Lakehouse.

SQL Is Always Part of Any Data Story

What a journey seeing what is possible using the power of Microsoft Fabric with Fabric Mirroring for SQL Server 2025. You can see it is far more than just copying your data into Fabric and running queries. Fabric Mirroring is just another example of how SQL is part of any data story for the industry, our customers, and Microsoft.

I asked Ajay Jagannathan, the lead program manager for Fabric Mirroring for SQL, the importance of Fabric Mirroring for the Microsoft data story: "*In today's AI-driven world, analytics platforms need to be near real time for quick insights and actions. Fabric Mirroring provides a modern way for customers to ingest their transactional data continuously into OneLake in Microsoft Fabric, thereby providing a solid foundation for reporting, advanced analytics, AI, and data science. Ever since we brought Azure SQL Mirroring to Fabric in 2024, customers have been asking us about bringing this capability to SQL Server. With a significant footprint of customers running SQL Server outside Azure, SQL Server 2025 provided us a perfect opportunity to bring the same Mirroring technology that powers Azure SQL to SQL Server. We have built seamless integration with Fabric ecosystem by allowing customers to centralize their SQL Server operational data into OneLake in an analytics-ready format using the same Fabric portal experience that is available for Azure SQL Database and Managed Instance. There is no complex setup or ETL needed for Mirroring. You set up the mirror from Fabric portal by providing the SQL Server and database connection details, provide selections on what needs mirrored into Fabric, either all tables or user-selected eligible mirrored tables. And just like that Mirroring is ready to go."*

After this chapter, it is fitting to end the book in the next chapter to talk about how SQL Server 2025 fits into the overall amazing story of SQL, ground to cloud to fabric.

CHAPTER 9

Microsoft SQL Ground to Cloud to Fabric

Up until 2024, I often would use the phrase **Microsoft SQL ground to cloud**. I would talk about *develop once, deploy anywhere* from SQL Server to Azure SQL. With the introduction of SQL Database in Fabric, I knew I needed to add in the third piece to the puzzle.

This chapter is short but sums up the incredible story of Microsoft SQL and SQL Server 2025.

Develop Once, Deploy Anywhere

Figure 9-1 represents the new develop once, deploy anywhere visual for Microsoft SQL, ground to cloud to fabric.

Note Microsoft SQL is not an official brand of Microsoft. It is my own personal tagline as there is no way in a single phrase to describe SQL Server, Azure SQL, and SQL Database in Fabric.

B. Ward, *SQL Server 2025 Unveiled*, https://doi.org/10.1007/979-8-8688-1847-9_9

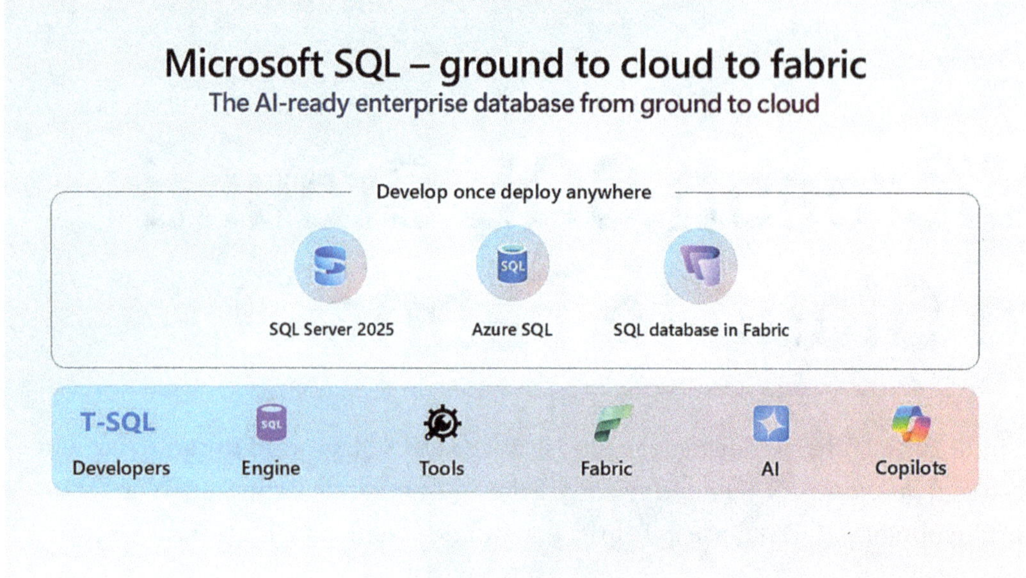
Microsoft SQL – ground to cloud to fabric
The AI-ready enterprise database from ground to cloud
Develop once deploy anywhere
SQL Server 2025
Azure SQL
SQL database in Fabric
T-SQL
Developers
Engine
Tools
Fabric
AI
Copilots

Figure 9-1. *Ground to cloud to fabric*

I've also used the statement "SQL is anywhere you need it," from SQL Server 2025 running on Windows or Linux, running in a container, or running within Kubernetes clusters to SQL Server in Azure Virtual Machines, Azure SQL Database, or Azure SQL Managed Instance and to now SQL Database running in Microsoft Fabric. It is clear Microsoft SQL exists in any form on any deployment platform you need.

SQL Server 2025

In this book you have learned the amazing story of SQL Server 2025, the AI-ready enterprise database. Remember my "camera slide" from Chapter 1 as you see in Figure 9-2.

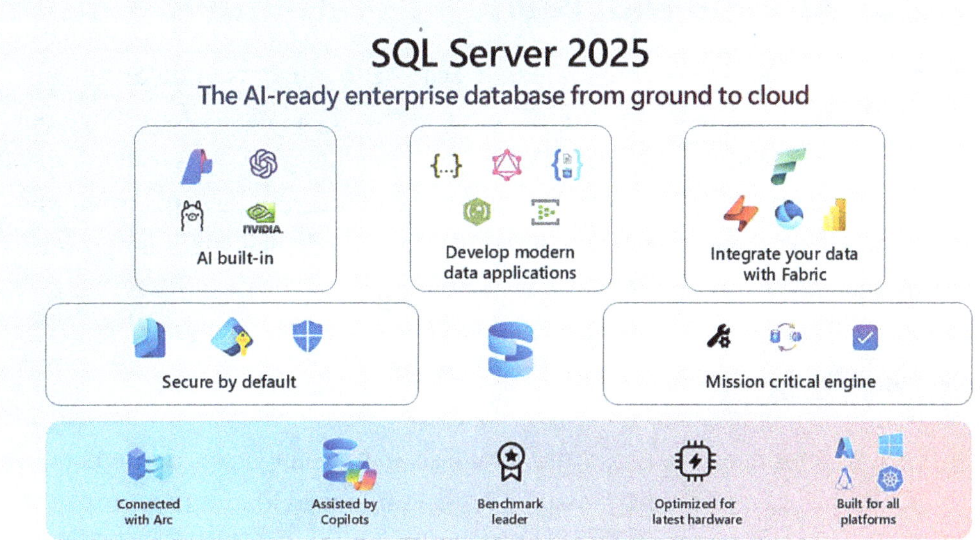

Figure 9-2. *A visual of SQL Server 2025 capabilities*

You have seen all the new capabilities from Chapter 1 all the way to Chapter 8 in the book for SQL Server 2025 including the following.

AI Built-In

In Chapter 4 you learned how to use AI models in a secure and isolated fashion to power a new form of searching called vector search. And you saw how to use these models using the security of SQL Server and with the T-SQL language using models that can run locally on-premises and in cloud providers. You also saw in Chapter 5 how to extend search results using REST APIs with **sp_invoke_external_rest_endpoint**, to provide rich recommendations from AI language models.

Developer Modern Data Applications

You saw in Chapter 5 of this book the rich set of developer features not seen in a decade. This includes native JSON support, Regular Expressions and other T-SQL enhancements, Change Event Streaming, and REST APIs including interactions with AI chat models. And don't forget to check out the two flavors now of Developer Edition: Standard and Enterprise.

Integrate Your Data with Fabric

In Chapter 8 you were able to see how valuable it can be to mirror your data with Microsoft Fabric, the unified data platform. Fabric Mirroring allows you to offload read workloads for near-real-time analytics, ETL free. This includes free storage in Microsoft Fabric for your mirrored data. I showed you how you can mirror data across multiple SQL Server sources and join them together in a single unified Lakehouse, empowering new analytic scenarios with Power BI and Data Agents.

Secure by Default

Security is not an option for any enterprise database. In Chapter 7, you learned how SQL Server 2025 brings new security features such as Managed Identity support with Microsoft Entra to go *passwordless*. This includes not just logins but *outbound* connections to destinations like Azure Storage and Azure AI Foundry.

New security capabilities for the core engine include security cache enhancements, new strong password and encryption algorithms, and TLS 1.3 support for features and tools.

Mission-Critical Engine

You also learned in Chapter 7 important performance enhancements including optimized locking, tempdb resource governance, intelligent query processing (IQP) enhancements, the query store on read replicas, and persistent stats on secondaries.

You also learned the amazing story of core enhancements for Always On Availability Groups including reliable and fast failover and tuning options. In addition, you learned about backup enhancements such as support for all backup types on secondaries, new backup compression options, and backups to immutable Azure Storage for ransomware protection.

You also saw some hidden gems in Chapter 7 including the ABORT_QUERY_EXECUTION query hint, new memory diagnostics in DMVs, and core performance enhancements in the database engine.

Foundations

SQL Server 2025 continues the story of Azure Arc-enabled SQL Servers including new Microsoft Entra Managed Identity support. New versions of SQL Server Management Studio, SSMS 21 and 22, are now available with Copilot assistance experiences.

SQL Server 2025 will set new performance benchmark standards when it becomes generally available and will include optimizations for the latest advancements in hardware (as you heard from Thierry Fevrier).

And remember that SQL Server 2025 runs on any platform you need including Windows and Linux, in containers, and in Kubernetes clusters, all using the same core database engine.

Azure SQL

In 2024, I launched the second edition of *Azure SQL Revealed*. In that book I showed Figure 9-3.

Azure SQL
SQL virtual machines
Managed instances
Databases
SQL
SQL
SQL
Azure manages the host and hardware
You manage the Virtual Machine
Azure provides value-add services
Azure manages the host, hardware, and Virtual Machine
You manage the SQL Server
Azure provides PaaS services
Azure manages the host, hardware, virtual machine, and SQL Server
You manage the database
Azure provides PaaS services

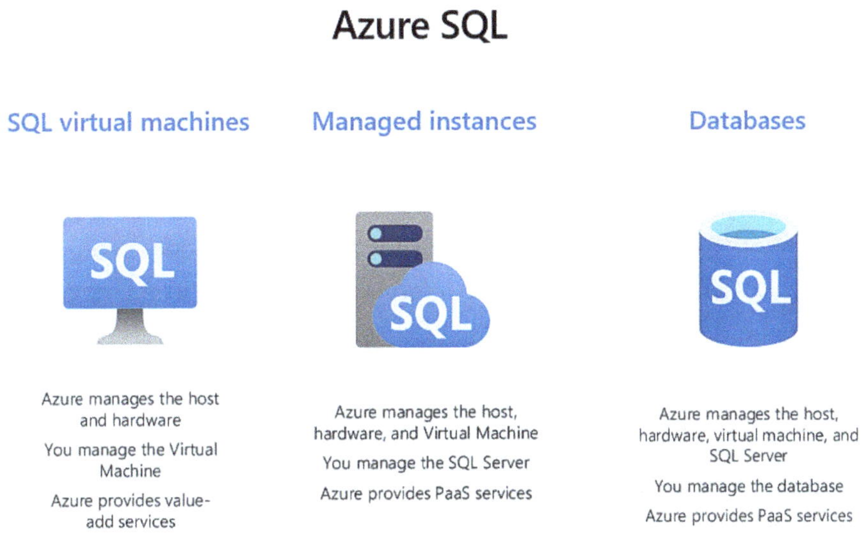

Figure 9-3. *The Azure SQL family*

In the book I showed you that Azure SQL is not a product but a *family* of options to deploy SQL Server in the Azure cloud from left to right.

SQL Server in Azure Virtual Machines

Think of this option as the full complete SQL Server "box" that you run on-premises in your own hosted virtual machine environment, but instead Microsoft provides the host. You deploy and configure the virtual machine using the Microsoft Infrastructure-as-a-Service (IaaS) Azure compute, storage, and networking platform. Microsoft will handle all host management including patching, resiliency to failures, and deployment options around the world.

Then you manage everything inside the VM including the operating system (Window or Linux) and SQL Server. Microsoft does provide some *managed* capabilities to help you including Microsoft Entra, monitoring, and update management. You also can deploy complex HA solutions with Always On Failover Cluster Instances and Availability Groups, optimized for the Azure cloud. You can learn more at `https://aka.ms/azuresqlvm`.

Azure SQL Managed Instance

With the VM story in mind, imagine that you don't have to manage anything about the VM itself or the OS *inside* the VM. You have access to a full SQL Server instance, which means multiple databases and SQL Server Agent jobs. But you don't have to worry about any aspect of the VM and OS including all update management.

In addition, SQL Server is versionless meaning Microsoft maintains and updates the SQL Server software itself. MI (that is our internal name for it) also includes automatic backups, automatic HA (with Availability Groups built-in), and deep integration with Microsoft Entra. Microsoft also has a new migration experience for Managed Instance using Azure Arc and Managed Instance Link.

If you are interested in MI, get started at `https://aka.ms/azuresqlmi`.

Azure SQL Database

At the far right of the managed spectrum is Azure SQL Database. This is the managed service that started it all back in 2008. Now imagine that you just manage a database. Microsoft manages the VM, OS, and SQL Server instance.

You have full control of the database, and you have all the managed capabilities of MI like versionless, backups, and HA.

The biggest advantage of SQLDB (our internal name) are the unique deployment options like serverless, elastic pools, and Hyperscale (allowing you to start small and autoscale to 30 replicas and 128TB of database size). I see us often innovate first in Azure SQL Database with our latest features. You can learn more at `https://aka.ms/azuresqldb`.

The farther left, the more control you have. The farther right, the more managed service capabilities you get.

All of these options have one thing in common: the same core SQL Server engine and T-SQL language. To get started looking at all your migration options for Azure SQL, see our documentation at `https://learn.microsoft.com/data-migration/sql-server/overview`.

SQL Database in Fabric

I briefly mentioned the amazing journey to launch SQL Database in Fabric in Chapter 8. You should think of SQL Database in Fabric as equivalent to Azure SQL Database but simplified, integrated, and optimized for the Microsoft Fabric platform.

Figure 9-4 shows the themes for SQL Database in Fabric.

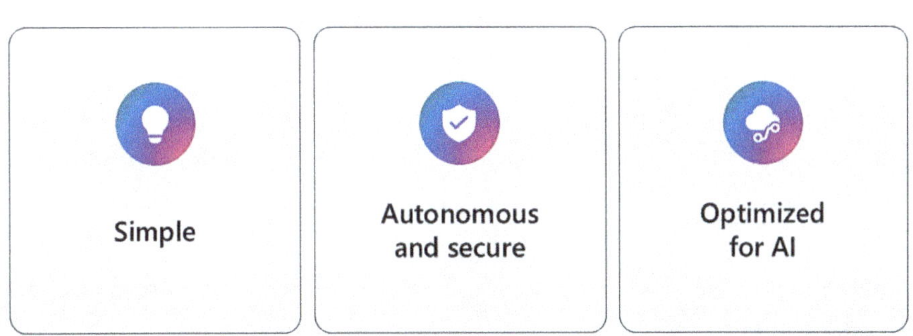

Figure 9-4. *SQL Database in Fabric*

Note One of the leading forces to start the SQL Database in Fabric project was Muazma Zahid. Then later the famous Anna Hoffman (from *Data Exposed* fame) took over product management (we needed Muazma to be awesome with our AI strategy, so why not both?). I know it was an incredible achievement for Anna to see this launch become public preview in November 2024 at Microsoft Ignite. I hope by the time this book comes out this product will become generally available. The more I think about it, someone should write a book about this product.

You can see from the themes of this project the idea of creating a Software-as-a-Service concept. This follows the themes of Microsoft Fabric, which started with a very simple database experience. This includes easy and fast deployment (just give it a name) and built-in experiences including all the automation and security that Azure SQL Database brings. But it also includes automatic Mirroring of your database into OneLake, built-in methods to generate GraphQL APIs, and built-in CI/CD pipelines to Git or Azure DevOps. One of the benefits of SQL Database in Fabric are the built-in integrations within Fabric including Mirroring, User Data Functions, Notebook support, and Power BI Translytical features.

In addition, like Azure SQL Database, we want developers to have a great database to build AI applications including built-in vector search, like you have seen with SQL Server 2025. Learn more at `https://aka.ms/sqldbfabric`.

A Common Bond

Some have said there are too many choices, and for many of the customers I talk to, they haven't settled on just one option. In other words, this is not an *or* but an *and* decision. Customers have come to use Microsoft SQL per their requirements in a mixture of these options.

How can you learn all of these options? The good news is the common bond that ties all of these together.

T-SQL

While you can find some differences, the T-SQL language is very common across all deployment options. Most of the time, when there is a difference, it is because the T-SQL statements may not make sense. For example, we don't offer any T-SQL to create or manage Always On Availability Groups for Azure SQL because Microsoft manages this capability.

Engine

Did you know the core database engine codebase is the same from ground to cloud to fabric? What happens is that we will sometimes accelerate code changes in the cloud first or only light up specific features from the core engine code where it makes sense. For example, we don't offer any specific Linux features for Azure SQL Database because you are abstracted from managing the OS. Slava Oks would always tell me, *"...the Query Processor is the Query Processor no matter where SQL lives."*

Tools

All our tools "just work" no matter where SQL Server is deployed whether it is SQL Server Management Studio or the mssql extension in Visual Studio Code. What is nice about our tools is how they adjust to the deployment choice you are connected to. Again, picking on Always On Availability Groups, SSMS won't offer any option for these for Azure SQL Database.

Fabric

We simply want you to have the ability to mirror SQL Server into Microsoft Fabric OneLake no matter where it lives:

- SQL Database in Fabric is mirrored by default.
- You learned in this book how SQL Server 2025 supports Mirroring.
- SQL Server 2016–2022 also support Mirroring.
- Azure SQL Database and Azure SQL Managed Instance also support Mirroring.

Learn everything about Fabric Mirroring for SQL at `https://aka.ms/sqlmirroring`.

AI

It is our intention that anywhere SQL lives, you can set up a vector search system using the AI model of your choice. SQL Server 2025 is ahead of the game, but Azure SQL and SQL Database in Fabric will catch up fast (they still support vector types and VECTOR_DISTANCE).

For our cloud options, we will pivot to support only Azure AI Foundry models (which can be more than just Azure OpenAI), but SQL Server in Azure Virtual Machines can support any AI model SQL Server 2025 supports.

In addition, **sp_invoke_external_rest_endpoint** works everywhere SQL lives. This means you can connect to any type of AI model inside the engine from a hosted service that supports REST. Azure options (except for VM) will support only Azure REST endpoints, including Azure AI Foundry.

Copilots

I am very excited to see the future of Copilot experiences in SQL Server Management Studio (SSMS). Since SSMS can connect anywhere SQL lives, you now have Copilot for any SQL.

Having said that, we also have GitHub Copilot integrated with the mssql extension for VSCode, the Azure Portal includes a Copilot experience, and the query editor in Microsoft Fabric connected to SQL and analytic endpoints has Copilot experience. Try these for yourself. A context-aware Copilot that can help you build and debug T-SQL is alone worth your time to explore.

The Future Is Bright

To some, inside and outside of Microsoft, it might feel that Microsoft SQL is taken for granted. It has been so successful for so long. Yet if you look at the innovation ground to cloud to fabric, SQL can still be so relevant for the most modern and innovation applications and business solutions.

I asked the famous Slava Oks, CTO and Distinguished Engineer at Microsoft (who also leads all the engineering of SQL ground to cloud to fabric) to summarize the importance of SQL: "*Azure SQL, SQL Server, and SQL Database in Fabric are everywhere data lives. Their ubiquity makes them foundational to both Microsoft and our customers. These crown jewel database offerings have evolved beyond mere databases into comprehensive data platforms that meet developers, DBAs, and database enthusiasts wherever they are—continually delighting everyone with new capabilities. In that spirit, the upcoming SQL Server 2025 release, the latest Azure SQL updates, and innovations in SQL Database in Fabric introduce game-changing advancements like a native vector data type, DiskANN-powered vector indexes for high-speed semantic search, hybrid search blending AI embeddings with traditional queries, T-SQL enhancements, integrated AI model management. These capabilities transform SQL Server, Azure SQL, and SQL Database in Fabric into foundational platforms for delivering on generative AI scenarios powered by Hybrid Retrieval Augmented Generation (Hybrid RAG), building intelligent assistants, and AI Agents. As a result, Azure SQL, SQL Server, and SQL Database in Fabric are strategically positioned at the heart of the Microsoft ecosystem to deliver on the promise of generative AI and to fulfill Microsoft's mission: 'to empower every person and every organization on the planet to achieve more.'"*

I hope you enjoyed learning everything about SQL Server 2025 in this book. I look forward to continuing to work for Microsoft to help shape the innovation of Microsoft SQL, ground to cloud to fabric.

Index

A

B. Ward, *SQL Server 2025 Unveiled*, https://doi.org/10.1007/979-8-8688-1847-9

D

E

F

G

H

I

J

K

L

M

N

O

P

Q

R

S

V

W, X, Y, Z

GPSR Compliance
The European Union's (EU) General Product Safety Regulation (GPSR) is a set of rules that requires consumer products to be safe and our obligations to ensure this.

If you have any concerns about our products, you can contact us on

ProductSafety@springernature.com

In case Publisher is established outside the EU, the EU authorized representative is:

Springer Nature Customer Service Center GmbH
Europaplatz 3
69115 Heidelberg, Germany

www.ingramcontent.com/pod-product-compliance
Lightning Source LLC
LaVergne TN
LVHW081347060526
838201LV00050B/1734

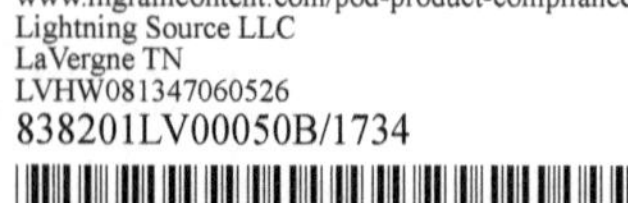
* 9 7 9 8 8 6 8 8 1 8 4 6 2 *